The Challenges to Nuclear Power in the Twenty-First Century

The Challenges to Nuclear Power in the Twenty-First Century

Edited by

Behram N. Kursunoglu

Global Foundation, Inc.
Coral Gables, Florida

Stephan L. Mintz

Florida International University
Miami, Florida

and

Arnold Perlmutter

University of Miami
Coral Gables, Florida

Kluwer Academic / Plenum Publishers
New York, Boston, Dordrecht, London, Moscow

Library of Congress Cataloging-in-Publication Data

The challenges to nuclear power in the twenty-first century / edited by Behram N. Kursunoglu, Stephan L. Mintz, and Arnold Perlmutter.
p. cm.
Includes bibliographical references and index.
ISBN 0-306-46491-8
1. Nuclear power plants--Risk assessment--Congresses. 2. Nuclear power plants--Safety measures--Congresses. 3. Nuclear power plants--Environmental aspects--Congresses. 4. Energy policy--Congresses. I. Kursunoglu, Behram, 1922- II. Mintz, Stephan L. III. Perlmutter, Arnold, 1928- IV. International Energy Forum (1999 : Washington, D.C.)

TK9152.16 .C485 2000
333.792'4--dc21

00-045616

Proceedings of the International Energy Forum 1999, held November 5–6, 1999, in Washington, D.C.

ISBN 0-306-46491-8

233 Spring Street, New York, N.Y. 10013

http://www.wkap.nl

10 9 8 7 6 5 4 3 2 1

A C.I.P. record for this book is available from the Library of Congress.

Printed in the United States of America

PREFACE

"International Energy Forum 1999" was held in Washington D.C. during November 5-6, 1999 in the Hyatt Regency Hotel in Crystal City. Once again the main topic was Nuclear Energy. Various papers presented contained pros and cons of Nuclear Energy for generating electricity.

We were aiming to clarify the often discussed subject matter of the virtues of Nuclear Energy with regard to Global Warming as compared to using fossil fuels for the generation of electricity. The latter is also currently the only way to operate our means of transportation like automobiles, planes etc. Therefore emission into the atmosphere of greenhouse gases constitutes the main source of Global Warming, which is absent in the case of Nuclear Energy.

These arguments are often put forward to promote the use of Nuclear Energy. However not all is well with the Nuclear Energy. There are the questions of the waste problem so far unsolved, safety of Nuclear Reactors is not guaranteed to the extent that they are inherently safe. If we aim to construct inherently safe reactors, then the economics of a Nuclear Reactor makes it unacceptable.

Year in and year out, we talk about these subjects but after the conference we do not do anything about it, we just go home. In the case of Nuclear Reactors the waste is localized in the place where the reactor is operating, while the waste generated by fossil fuels is spread globally. In other words we have here an irreversible hazard that is entirely out of our hands.

The participants of this conference and of the future ones, will continue thinking and discussing these subjects and perhaps we shall eventually overcome these difficulties. At present we do not know the answers but we hope for the best.

The Chairman and Trustees of the Global Foundation wish to gratefully acknowledge the generous support of this conference by The Electric Power Research Institute, Palo Alto, California, Nuclear Energy Institute, Washington D.C. and USEC, Inc., Bethesda, Maryland.

Behram N. Kursunoglu
Stephan L. Mintz
Arnold Perlmutter
Coral Gables, Florida
February 2000

About the Global Foundation, Inc.

The Global Foundation, Inc., which was established in 1977, utilizes the world's most important resource... people. The Foundation consists of distinguished men and women of science and learning, and of outstanding achievers and entrepreneurs from industry, governments, and international organizations, along with promising and enthusiastic young people. These people convene to form a unique and distinguished interdisciplinary entity to address global issues requiring global solutions and to work on the frontier problems of science.

Global Foundation Board of Trustees

Former Trustees

GLOBAL FOUNDATION'S RECENT CONFERENCE PROCEEDINGS

Making the Market Right for the Efficient Use of Energy
Edited by: Behram N. Kursunoglu
Nova Science Publishers, Inc., New York, 1992

Unified Symmetry in the Small and in the Large
Edited by: Behram N. Kursunoglu, and Arnold Perlmutter
Nova Science Publishers, Inc., New York, 1993

Unified Symmetry in the Small and in the Large - 1
Edited by: Behram N. Kursunoglu, Stephen Mintz, and Arnold Perlmutter
Plenum Press, 1994

Unified Symmetry in the Small and in the Large - 2
Edited by: Behram N. Kursunoglu, Stephen Mintz, and Arnold Perlmutter
Plenum Press, 1995

Global Energy Demand in Transition: The New Role of Electricity
Edited by: Behram N. Kursunoglu, Stephen Mintz, and Arnold Perlmutter
Plenum Press, 1996

Economics and Politics of Energy
Edited by: Behram N. Kursunoglu, Stephen Mintz, and Arnold Perlmutter
Plenum Press, 1996

Neutrino Mass, Dark Matter, Gravitational Waves, Condensation of Atoms and Monopoles, Light Cone Quantization
Edited by: Behram N. Kursunoglu, Stephen Mintz, and Arnold Perlmutter
Plenum Press, 1996

Technology for the Global Economic, Environmental Survival and Prosperity
Edited by: Behram N. Kursunoglu, Stephen Mintz, and Arnold Perlmutter
Plenum Press, 1997

25th Coral Gables Conference on High Energy Physics and Cosmology

Edited by: Behram N. Kursunoglu, Stephen Mintz, and Arnold Perlmutter
Plenum Press, 1997

Environment and Nuclear Energy
Edited by: Behram N. Kursunoglu, Stephan Mintz, and Arnold Perlmutter
Plenum Press, 1998

Physics of Mass
Edited by: Behram N. Kursunoglu, Stephan Mintz, and Arnold Perlmutter
Plenum Press, 1999

Preparing the Ground for Renewal of Nuclear Power
Edited by: Behram N. Kursunoglu, Stephan Mintz, and Arnold Perlmutter
Plenum Press, 1999

Confluence of Cosmology, Massive Neutrinos, Elementary Particles & Gravitation
Edited by: Behram N. Kursunoglu, Stephan Mintz, and Arnold Perlmutter
Plenum Press, 1999

International Energy Forum 1999
Edited by: Behram N. Kursunoglu, Stephan Mintz and Arnold Perlmutter
Plenum Press, 2000

International Conference on Orbis Scientiae 1999
Quantum Gravity, Generalized Theory of Gravitation and Superstring Theory-based Unification
Edited by: Behram N. Kursunoglu, Stephan Mintz, and Arnold Perlmutter
Plenum Press, 2000

Global Foundation, Inc.

A Nonprofit Organization for Global Issues Requiring Global Solutions, and forProblems on the Frontiers of Science

Center for Theoretical Studies

INTERNATIONAL ENERGY FORUM 1999
(22nd In A Series of Conferences Since 1974)

November 5 - 6, 1999

Hyatt Regency Hotel
Crystal City

Meeting
Roosevelt Room

This conference is supported in part by

Electric Power Research Institute,
Nuclear Energy Institute ,
US EC Inc.

Sponsored by:
Global Foundation Inc.

P. O. Box 249055
Coral Gables, Florida 33124-9055
Phone: (305) 669-9411
Fax: (305) 669-9464
E-mail: kursun@globalfoundationinc.org
Web: http://www.globalfoundationinc.org

Conference Hotel:
Hyatt Regency Crystal City

At Washington National Airport
2799 Jefferson Davis Highway
Arlington, VA 22202
Phone: 800-233-1234
Fax: 703-418-1233
Group rate: $145.00/night
Single/Double occupancy

DEDICATION

The trustees of the Global Foundation and members of the 22nd Energy Conference, dedicate this conference to Dr. Glenn T. Seaborg of Lawrence Berkeley Laboratory at the University of California at Berkeley. The late Professor Seaborg was a loyal and active member of this series of conferences on energy issues since 1974. He served as a trustee of the Global Foundation. Dr. Seaborg was awarded a Nobel Prize in Chemistry for his discovery of various trans-uranium elements, one of which was named Seaborgium in his honor. He also served for ten years as Head of the United States Atomic Energy Commission. His presence at the University of California at Berkeley helped greatly in increasing the volume and quality of scientific research there. We shall all miss Glenn. We extend our deepest condolences to his family.

--NOTES--

1. Each presentation is allotted a maximum of 30 minutes and an additional 5 minutes for questions and answers.
2. Moderators are requested not to exceed the time allotted for their sessions.

Moderator: Presides over a session. Delivers a paper in own session, if desired, or makes general opening remarks.

Dissertator: Presents a paper and submits it for publication in the conference proceedings at the conclusion of the conference.

Annotator: Comments on the dissertator's presentation or asks questions about same upon invitation by the moderator.

CONFERENCE PROCEEDINGS

1. The conference portfolio given to you at registration contains instructions to the authors from the publisher for preparing typescripts for the conference proceedings.

2. Papers must be received at the Global Foundation by January 8, 2000.

3. An edited Conference Proceedings will be submitted to the Publisher by February 15, 2000.

INTERNATIONAL ENERGY FORUM 1999
PROGRAM

Friday, November 5, 1999

8:00 AM Registration

8:30 AM Opening Session: **Behram N. Kursunoglu**
Chairman of the Global Foundation, Inc.
Opening Remarks

Keynote Dissertator: **Nils J. Diaz**, NRC, Washington, D.C.
"Regulation, Reliability and Competition"

Keynote Address: **Ernest J. Moniz,**
Under Secretary of Energy
"Shaping the Energy Future"

10:00 AM Coffee Break

10:30 AM SESSION I. **Energy: An Ecumenical View**
Why/ whether policy is needed: demand needs and reality; decarbonization of world metabolic cycle; general background history; role of energy in development ,etc.

Moderator: **Edward Arthur,** Los Alamos National Laboratory

Dissertators: **Herb Inhaber,** University of Nevada
"Creating a level playing field in the public mind: How to incorporate energy risks and environmental effects"

Peter Beck, Royal Institute of International Affairs
"The challenges to Nuclear Power In the Next Century: can they be overcome?"

Thomas Blejwas, Sandia National Laboratories
"The Emerging Nuclear Future and National Security"

Annotator: **Juan Eibenschutz,** Luz y Fuerza del Centro, Mexico

Session Organizer: **Edward Arthu**

12:00 Noon: Lunch Break

1:00 PM : SESSION II: EUROPEAN PERSPECTIVE

Moderator: **Gerald Clark**, Uranium Institute

Dissertators: **Werner Süess,** Münich
"The future of Nuclear In a Competitive Market"

Andre Lacroix, Electricité de France
"EdF and Liberalisation of The Market"

Annotator: **Pierre Zaleski**, Université Paris Dauphine

Session Organizer: **Gerald Clark**

1:30 PM SESSION III: **Need For Nuclear Energy**

Moderator: **Bertram Wolfe,** GE Nuclear

Dissertators: **Robin Jones,** Electric Power Research Institute
"Power Generation Diversity: A Global Imperative"

Angie Howard, Nuclear Energy Institute, Washington D.C
"Convergence Of Favorable Thinking about Nuclear Energy"

3:00 PM Coffee Break

3:30 PM **Clinton Bastin,** Formerly US Department of Energy
"Nuclear Technology: Need For New Vision"

David Bodansky, Professor Emeritus, University of Washington
"Nuclear Power in the context Of Critical Global Problems"

Annotators: **A. David Rossin,** Former Assistant Secretary For Nuclear Energy, US Department Of Energy

Craig F. Smith, Lawrence Livermore Nat'l Lab

Session Organizer: Bertram **Wolfe**

4:30 PM SESSION IV: **Public Acceptance for Nuclear Energy**

Moderator: **John R. Ireland,** Los Alamos National Laboratory

Dissertators: **Stefan Hirschberg,** Paul Scherrer Institute
"Combining Technical Knowledge and value Judgments to Guide Decisions"

Annoteter: **Bertrand Vieillard- Baron, Framatome**

Session Organizer: **Robert Krakowski**

6:00 PM *Conference adjourns for the day*

7:30 PM Conference Banquet, Tidewater Room
P.M. WIHBEY, After Dinner Speaker

Saturday, November 6, 1999

8:30 AM Session V: **Obstacles To A Level Playing Field**
Evaluations/ Assesments

Moderator: **Myron Kratzer,** Washington D.C.

Dissertators: **William H. Timbers,** USEC Inc.
"Why we must remove the Roadblocks To More Nuclear Power"

John G. Strand, Michigan Public Service Commission
"Nuclear Power and Electric Industry Restructuring"

10:00 AM Coffee Break

10:30 AM **Juan Eibenschutz** , Luz y Fuerza del Centro, Mexico
"The Confusion Between Constraints and Objectives"

James Tape, Los Alamos National Laboratory
"Commercial Nuclear Power and Proliferation: What is Proliferation Resistance? "

Annotators: **Angie Howard**

Janice E. Owens, Edlow International Company, Washington D.C.

Session Organizer: **Myron Kratzer**

12:00 Noon Lunch Break

1:00 PM **Concluding Panel Discussion**

Moderator: **Gerald Clark**
Panel Members: **Peter Beck , Juan Eibenschutz, Myron Kratzer, William Timbers, Bertram Wolfe, Pierre Zaleski**

3:00 PM *1999 Global Foundation Energy Conference Adjourns*

CONTENTS

SECTION IV
Concluding Panel Discussions

SECTION V
After Dinner Speech

The Challenges to Nuclear Power in the Twenty-First Century

THE PROBLEM OF ENERGY AND NUCLEAR MATTERS*

Behram N. Kursunoglu

Global Foundation, Inc.

*Excerpt from "The Ascent of Gravity" the author's typescript to be published.

1. ESTABLISHMENT OF THE GLOBAL FOUNDATION INC.

The 1970s began with continuing global challenges, the worsening of the cold war between the USSR and the West, the energy crisis that began with the embargo imposed by the petroleum exporting countries in 1973, ever increasing world population, the proliferation of nuclear weapons, the increasing gap between the rich and poor, global environmental degradation and more. These issues deeply motivated my colleagues and myself to participate in the analysis and understanding of these fundamental issues. I saw a great opportunity to bring together distinguished people under the ægis of the Global Foundation. It was established in 1977 and incorporated in 1978 as a not-for-profit corporation under the laws of Florida and was granted a 501(c)(3) tax-exempt status by the Internal Revenue Service.

The Foundation would consist of great men and women of science and learning and of outstanding achievers and entrepreneurs from industry, government, and international organizations, along with promising and enthusiastic young people. These people would form a unique and distinguished, interdisciplinary, intellectual expertise and the Foundation would be dedicated to assembling the resources necessary for them to work together. Beginning in 1992, the Center for Theoretical Studies became part of the Global Foundation and thus the work of the Foundation would include global issues and frontier problems in science. The Foundation's work, therefore, is a common effort, employing the ideas of creative thinkers with a wide range of experience and viewpoints.

I discussed these ideas in detail with the late Robert Hofstadter (Stanford University) in 1977 who accepted to serve as one of the trustees of the Foundation. The remaining trustees included Willis E. Lamb, (Yale University), Glenn T. Seaborg,

The Challenges to Nuclear Power in the Twenty-First Century
Edited by Kursunoglu *et al.*, Kluwer Academic/Plenum Publishers, New York, 2000

(University of California, Berkeley), Abdus Salam (International Center for Theoretical Physics, Trieste), Frederick Reines (University of California at Irvine), Louis Néel (University de Grenoble), Eugene P. Wigner (Princeton University), and Manfred Eigen (Max-Planck Institut, Göttingen). These Nobel Laureates were joined by other distinguished people: Jean Couture (Institute Français de L'Energie, Paris and former Secretary of Energy for France), Henry King Stanford, (President Emeritus of the Universities of Miami and Georgia), and the late Lord Solly Zuckerman, OM (University of East Anglia, London).

The most recent additions to the Global Foundation trustees were Lord Marshall of Goring of London whom I met in 1961 at the Harwell British Atomic Energy Establishment where he was the head of its theoretical physics division, and Robert Herman (University of Texas, Austin) who was a longtime friend of Hofstadter and who was one of the physicists that recommended Herman for the Nobel prize. Herman's work with Ralph A. Alpher on the 3 degrees Kelvin microwave radiation left over from the Big-Bang was hailed by physicists and cosmologists. I was also among those physicists who recommended him along with Alpher for the prize to the Nobel Foundation.

In fact, the four most important advances in cosmology in the past five decades after Hubble's observation of the expanding universe, and after George Gamow's Big-Bang theory, include in chronological order: (1) Ralph A. Alpher and Robert Herman's theoretical prediction in the 1940s of the cosmic microwave background radiation (CMBR) left over from the Big-Bang (2) Arno A. Penzias and Robert W. Wilson's observation in 1964 of the residual heat detected as the CMBR which brought to these two members of the Bell Telephone Laboratories the Nobel prize; (3) Alan H. Guth's hypothesis in 1979 of an "inflationary" universe, (4) the observation in 1992 of microwave anistropies in the CMBR as seen through COBE (NASA's Cosmic Background Explorer) by physicists and cosmologists led by George Smoot. Unfortunately, both Lord Marshall and Robert Herman passed away during 1996 and 1997, respectively. The year 1997 was an unlucky one for the Global Foundation during which we also lost Abdus Salam.

The relationship between cosmology and elementary particle physics is one of the frontier fields currently pursued by the Global Foundation. There is amongst physicists and cosmologists a consensus that a unified theory of the large and the small (cosmology and elementary particles) is essential to a complete description of either. Foundation conferences are an opportunity for scientists to present and discuss their research and theories towards such a unification.

The real value of the Foundation to world society depends on the degree to which it succeeds in meeting its global purpose. To that end, the Foundation has published over 24 books on a variety of topics such as cosmology, elementary particle physics, environment, nuclear energy and technology. Many of these books are the fruits of the conferences that bring great minds together under the auspices of the Global Foundation.

Through its own activities and its participation in and support of institutions that are interested in the same issues and problems, the Global Foundation focuses its research, education, and training programs on the following issues which exacerbate global problems:

The Nuclear Quagmire

Nuclear war and nuclear peace are fundamental concerns of humankind. The Foundation concentrates on the strategic, technological, scientific, and political analyses of this issue in an interdisciplinary atmosphere.

The Foundation's seminars and workshops include the following topics: proliferation and control of weapons of mass destruction and delivery systems by the U.S. and the Russian Federation; the possible role of proliferation by the unemployed weapons scientists in Russia; dismantlement of nuclear, chemical, and biological weapons by the U.S. and Russia; non-proliferation roles of the EU, Japan, and NATO; export restrictions and controls--NATO, IAEA, and United Nations roles in global missile defense; activation of GPALS (Global Protection Against Limited Strikes); and the choices of technology transfer restrictions and preemption of global missile defense through the United Nations.

International Environmental Problems

The Foundation seeks solutions to global, international, and regional environmental problems through research, seminars, conferences and forums. The responsibility for environmental protection must be shared by all inhabitants of our planet. The most powerful means of addressing this greatest of all global issues is worldwide education. The Foundation created a program of higher education in developed and developing countries by organizing workshops, entitled "Education of Sustaining a Livable World Environment: Energy, Population, and Resources," for university professors who would be teaching courses on the subject. Included among the workshop lecturers were speakers from the countries represented by the participants.

The one-week, intensive workshop curriculum included scientific, technical, and economic analysis of subjects in the fields of atmosphere and climate evolution, global warming; decreasing biodiversity, and habitats in the Amazon and other regions with tropical rain forests; environmentally wise manufacturing and sustainable development; the uncertainty between sustaining a livable world and population growth; more efficient and environmentally compatible uses of energy, nuclear power and the environment; and the pros and cons of nuclear power for developing countries. The lectures were delivered by world-class scholars. Factual and observational data were presented, and all the lectures were followed by discussion periods. The evening sessions included film presentations on the Amazon, nuclear power, fossil fuels and the environment, the polluted Mediterranean Sea, the state of the Antarctic and the Arctic, and population.

Problems of Energy

The Global Foundation organizes conferences on the problems of energy as they relate to industrially developed and developing countries' needs. We study the energy interdependence of these countries and recommend appropriate sources of energy which are compatible with environmental protection, economic development, and the roles of fossil fuels, solar energy, nuclear energy, hydroelectric dams, etc.

The Foundation also engages in theoretical research on the following problems:

1) Elementary particles and the nature of the universe (including studies on the general theory of relativity and its generalizations)

2) The origin of the Solar System

3) Evolution of the genetic code

4) Physics and chemistry of memory and neural membranes

5) New forms of matter

2. ENERGY BLACK HOLE

"The Energy Black Hole" was the title of my presentation in our conference "Topics in Energy and Resources" held in January, 1974, in Coral Gables. It was 20 years ago when China had a population of 750 million compared to 1.25 billion in the China of today. The Indian population stood at 600 million, versus one billion now, and in 1974 India had just produced its first atomic bomb. Recently, in 1998, both India and Pakistan actually detonated hydrogen bombs, thereby becoming members of the nuclear club. The two countries may now find it easier to talk to each other than when only one of them had the bomb.

The conference was motivated by the embargo imposed by oil-exporting Arab countries in 1973. It sent a shock wave through western countries and created long lines of cars waiting to gas up at service stations all over the United States and Europe. The over-all mood was quite pessimistic and oil became the most precious commodity. Speculations on the "already" exhausted oil and natural gas reserves were being made with a vengeance. At the time, total world consumption of oil was about 60 million barrels per day of which 15 million barrels per day were being used by the U.S. alone. However, none of these pessimistic forecasts were true.

The current proven world reserves (i.e., those reserves that are economically and technically feasible to develop at today's prices) consist of one trillion barrels of oil, four quadrillion cubic feet of natural gas, and the coal reserves in China, Russia, and the U.S. are equivalent to an energy content exceeding that of total oil and natural gas reserves. The numbers keep changing from year to year but they mostly favor the large reserves of the earlier estimates. China's Tarim Basin is one of the most inhospitable regions of the world. The people who dwell on its edge describe it as a place where "you can get in, but not out." Yet within the region are potential oil reserves of 270 to 300 billion barrels.

Thus, the Tarim Basin in China's far west is like a black hole such that what falls in remains there with no way out. Several geological surveyors of the region have been sucked in without leaving a trace. At least 34 people have died during the fierce sandstorms. The Tarim is known to be the largest unexplored oil basin in the world and is a great frontier for a major source of energy. Unfortunately extractions of oil from the region could come only at a very high price. At present, 75% of China's energy needs are met with coal which produces large amounts of waste in the form of greenhouse gases like sulfur dioxide.

A cycle of global climate changes is often predicted to occur over the next few decades, which will be unfavorable to agricultural production. Thus, we may face not only an increasing demand for food for survival, but the specter of global famine leading to deaths of hundreds of millions of people. Under such extreme circumstances, a crisis of water shortages may necessitate the desalination of sea water in order to grow subsistence crops. It is estimated that it will take about 50 megajoules of energy to desalinate one cubic meter of sea water. On this basis we have to double the current world energy consumption to meet a subsistence level of agriculture. However, if we include the technological, economic, and industrial layout necessary and the corresponding strain on other life-sustaining uses of materials, then we may be facing a task that we have never had to undertake for survival. This means that we may reach a critical point where energy, usable without endangering the environment, is considerably less than the required amount for the continuation of life on this planet. Are we going to force a resource management crisis? Under these circumstances, it would take a short time, compared to the time spanned by the entire history of our civilization, for life, under its own weight, to collapse. Thus, the earth would become a dead planet or an "*energy black hole*." In order to avoid or to indefinitely postpone such an eventuality, we must first recognize the global nature of the problem. The energy interdependence of nations must not be based on the geopolitical distribution of resources, know-how, technological capabilities and potentials alone. All of these pessimistic forecasts may sound like wise or unwise cracks that have been said before, but if we wait "long enough" it may be too late to do anything. The uncertainty lies in the time span of things to come.

The rate of population growth is, undoubtedly, the greatest threat to survival regardless of where it occurs, in developing nations or in technologically advanced nations, they will strain equally the resources of the world. These facts demonstrate clearly that no nation, neither rich nor poor has any acceptable reasons for *free population growth*. In fact, the "zero population growth" often advanced as an ultimate solution to most of the frontier global problems (energy, environmental deterioration, hunger, transnational migrations, development, proliferation of weapons of mass destruction) is not enough. Mankind's chances for survival are, beyond a certain level, inversely proportional to the number of people on earth and directly proportional to the judicious and ingenious use of resources. A world-wide "negative growth" of population from the current 6.5 billion to 3 to 4 billion people must be adopted by all nations as the ultimate goal to reach a state of global equilibrium to avoid total collapse of our civilization.

The modernization of agriculture and the optimization of agricultural resources is not enough; it must be paired with negative population growth. World organizations, such as the United Nations and NATO that were established for the solution of a multitude of problems, must extend their efforts beyond peace keeping to the resolution of the problems at the root of conflict. The time has come for these organizations to adopt new principles and goals, compatible with present and forthcoming changes. The United Nations' mission when it was established in 1945 was, quite naturally, to focus all its efforts on the maintenance of world peace, prevention of war without sufficient focus on the real causes of war and peace. The problem is not how to feed, clad, provide shelter, transport, educate and supply technology to a growing world population. The real problem is growth with quality which is strongly dependent on the negative growth of world population to a steady state level of perhaps 3 billion people. It is, certainly, preferable that 3 billion people exist happily and with dignity, than it is to have the current 5.5 billion or, with unchecked growth, the expected 10 billion and beyond, live in the "energy black hole" described above.

Thus, real interdependence of nations from the point of view imposed by the greatest problem of all time, energy, is not in the distribution of energy from the resource-rich nations to resource-poor nations, but in the ultimate increase of per capita energy consumption in a world containing a population much less than its current levels. This proposal is, perhaps, not a practical one but it is the only sane path. Unfortunately, not every sane path can be trod. The alternative, namely unspared efforts to try all possible roads to sustain a growing world with finite resources is in the short run a practical approach but its end result in the not too distant future is the "energy black hole." We may of course, question the premise of having only three billion happy and prosperous people in the world instead of a world of free growth population with very few prosperous and mostly hungry and homeless human beings who would prefer life as it comes. The latter alternative is the likely future of *Homo sapiens* on this planet.

The old and still existing myth, the greater the population of a country the better for its defense against foreign aggression is no longer true. In fact, in proportion to its resources and land, a country with a smaller population could socially and economically be more stable than a country with a much larger population. For example, two hundred million persons in the Indian subcontinent without an atom bomb, would be economically a much more viable society and militarily a much stronger one, than would a society with one billion unsustainable members in possession of atomic weapons. The latter ought to apply to all nations in possession of atomic weapons: ban the bomb.

The greatest aid to developing countries with rapidly increasing population from the rich nations can consist of setting direct examples in population control. The NATO countries, as a military and political bloc, could begin by initiating a negative population growth plan in their own countries. In fact, because of the enormously wasteful economic structures of these countries, their population growth is the most serious in the world. In NATO countries the per capita strain on environment and global resources are among the greatest in the world. The consequences of "negative population growth" are not known but in case of an undesirable outcome the process can be reversed. We may find that a pulsating global population of between 3 and 4 billion people is the most realistic solution to the problem of resources, energy, and environment.

3. THE ENERGY FORUM OF 1977

At this point it is relevant to briefly remind the reader that the problem of nuclear energy was one of the earliest sources discussed in detail. The following White Paper was the first one issued by the Foundation.

White Paper, 1977 Energy Forum

Preface

The newly established Global Foundation and the Center for Theoretical Studies began a series of international energy forums to be held in various countries throughout the world. The aims are to bring together transdiciplinary, transnational, high level experts, along with younger people interested in energy problems. The forums are designed to address themselves to specific opinion and governmental policies as may be effected by such a meeting. A White Paper expressing a consensus of opinion of its signatories is to be issued at the end of each forum.

White Paper

On the occasion of the International Scientific Forum on An Acceptable Nuclear Energy Future of the World, held in Fort Lauderdale, Florida, from November 7 through 11, 1977, and sponsored by the newly established Global Foundation and the University of Miami's Center for Theoretical Studies, the undersigned have considered global energy requirements for the future and, also, world developments to meet this demand.

It was generally agreed that:

1. *World demand for energy will increase strongly as the standard of living and the size of presently disadvantaged populations increase over the next several decades.*
2. *Failure to meet this demand will result in extensive social evils such as poverty, starvation, unrest, epidemics, riots, and wars.*
3. *No single technology can meet the world's future demand. It is likely that all technologies, such as conventional fossil, nuclear fission, nuclear fusion, geothermal, and solar technologies, will be required to meet the qualitative and quantitative aspects of the demand, just as today no single technology meets all demands.*
4. *Nuclear fission must play a significant role in meeting world demand over the next several decades, and over this period, its full exploitation cannot be foregone without excessive risk.*
5. *An assured nuclear fuel supply, of utmost importance to many nations, cannot be guaranteed by uranium mining alone. Although the urgency will vary from country to country, in the application of nuclear fission energy, fuel reprocessing is essential. Further, the best way to handle spent fuel and to take care of nuclear wastes is to reprocess the spent fuel.*
6. *There are many candidate systems that may be called upon to supplement or, eventually, to replace our present largely light water reactor technology. These include fast breeder reactors. Developments in these systems should be pursued vigorously on an international basis, although not necessarily all systems in all countries.*
7. *Practical consideration of the ability to produce and deploy reactors in the numbers necessary dictates that currently successful systems be sustained and their installation encouraged by governments until and unless advanced systems are fully available and acceptable technically, economically, and industrially.*
8. *The plutonium-uranium fuel cycle has particular advantages in fast spectrum reactors and the uranium 233-thorium fuel cycle in thermal reactors. Both will need to be developed, including all necessary steps for full implementation.*

9. *Impressive progress has been achieved toward proving the scientific feasibility of fusion systems based on the principles of magnetic and inertial confinement. Progress has been made also that suggests systems that, on a longer time scale, may indicate economic feasibility. Development of these systems, already involving a considerable degree of international cooperation, should be pursued vigorously on this basis; again, not all systems in all countries. However, the possible successful development of fusion technology should not delay the prudent and necessary deployment of fission technology. It is possible that the first application of fusion technology will be in a hybrid fission-fusion system.*

10. *It is recognized that the deployment of fission power or hybrid fusion-fission power on a large scale poses problems of safeguards of material against potential diversion and, thus, proliferation of nuclear weapons. We are confident that the international community can and should take the political, institutional, and technical measures that will be effective in diminishing the risk of proliferation, while retaining the economic advantages of nuclear power. Therefore, we do not believe that the risk of proliferation should deter the use of nuclear energy.*

11. *The probability that accidents in existing reactors will cause harm is acceptably small, and we believe with proper use of experience will diminish even as the number of reactors increases.*

12. *Solar energy may have a part in the mixed energy system of the future. The extent of its penetration will depend largely upon economic considerations. It is difficult to determine finally what these economic parameters will be without practical experience on a substantial scale; at present parameters appear to be adverse.*

13. *Meeting the energy demand of the still rapidly increasing world population with legitimate expectations of a higher standard of living calls for large scale mobilization of labor, materials, capital, and technical and management skills. It should be the constant preoccupation of governments to accomplish this economically and effectively to avoid overtaxing the world's population capabilities and resources of these necessities.*

14. *There is an urgency to the world energy problem that, especially in view of the long lead times, brooks no delay in determining and executing national programs and in seeking international cooperation to take up the tasks and share the benefits equitably.*

Nikolai G. Basov, P.N. Lebedev Physical Institute, USSR Academy of Sciences Moscow, USSR*

Hans A. Bethe, Cornell University, Ithaca, New York*

Karl Cohen, General Electric Company, San Jose, California

Floyd Culler, Oak Ridge National Laboratory, Oak Ridge, Tennessee

Robert Hofstadter, Stanford University, Stanford, California*

Behram N. Kursunoglu, Center for Theoretical Studies, University of Miami, Coral Gables, Florida

W. Bennett Lewis, Queen's University, Ontario, Canada

Marjorie P. Meinel, University of Arizona, Tucson, Arizona

Keichi Oshima, University of Tokyo, Tokyo, Japan

Edward Teller, Stanford University, Stanford, California

Alvin Weinberg, Institute for Energy Analysis, Oak Ridge Associate Universities, Oak Ridge, Tennessee

Eugene P. Wigner, Princeton University, Princeton, New Jersey*

Pierre Zaleski, Embassy of France, Washington, D.C.

Edwin Zebroski, Electric Power Research Institute, Palo Alto, California

**Nobel Laureate*

----End of White Paper----

4. THE VIEWS FROM THE CAPITOL

The first International Scientific Forum on Energy in November, 1977 was subtitled "Nuclear Energy Future of the World" and was held in Fort Lauderdale, Florida. The white paper issued on this occasion had favorable recommendations on nuclear energy and the cosigners of this declaration included some of the biggest names of the nuclear era.

The 1977 forum was followed in 1978 on the same subject bearing the subtitle "Nuclear Energy Alternatives." However, the Three Mile Island accident in March of 1979 was a serious blow to the safety credibility of nuclear energy as a source of electricity generation. After conferring with a few members of the planning committee for the energy fora (Hans Bethe, Robert Hofstadter, Eugene P. Wigner, and Edward Teller), we decided to make a presentation on nuclear energy before the House Committee on Science and Technology. The congressional part of the proposed meeting was, at my request, organized by Congressman Dante Fascell from South Florida.

We all arrived in Washington, D.C. on May 6, 1979, and I called a rehearsal meeting in the evening to organize our presentation. The presentation before the Congressional Committee was on May 7, 1979, at 2:00 p.m. in room 2318, Rayburn House Office Building, the Honorable Don Fuqua, chairman of the Committee, presiding. It was agreed in our previous evening's meeting in the hotel that I should call on the members of our committee to make their presentations. Chairman Don Fuqua invited Congressman Carl Pursell to introduce our panel. He thanked me and elaborated on the introduction of each of us and especially that of Teller and the Nobel laureates Bethe, Hofstadter, and Wigner. I was the first speaker asked to describe our mission by giving a brief report on our white paper favoring, despite the Three Mile Island accident, nuclear energy as a major generator of electricity. The following material was included in the congressional hearings as recorded by Regina A. Davis, Chief Clerk. I was invited by the chairman to begin the testimony which I started by reading our 1977 white paper to sum up the panel's viewpoints on energy.

The continuing disagreement between Teller and Hofstadter regarding their divergence of opinion on fusion energy was, unfortunately, not contained and came out in their testimonies despite promises by both gentlemen the previous evening to be discreet. In fact, Edward Teller's deep-seated views against realization of fusion reactors and the contrary views of Hofstadter were claimed by Edward Teller as the cause for his heart attack upon return, after the meeting, to San Francisco! These disagreements were never ironed out between the two distinguished physicists. There were, at the meetings, questions, answers and comments on nuclear energy, energy conservation, fossil fuels, energy needs of developing countries and the non-competitive status of solar energy. The panel regarded the Three Mile Island accident as a very serious combination of system and human failures. A sigh of relief came from the fact that there was not a melt-down of the reactor core. The adverse effect was psychic damage on the public. The Three Mile Island accident was an ideal opportunity for panic-mongers to try to discredit nuclear energy and reeducate the public against it.

When I called on Eugene Wigner to discuss the subject of waste disposal, fuel reprocessing, and proliferation problems, he began by remarking that "the Three Mile Island accident was really a positive one, because in spite of all the mistakes that have been made, and in spite of, also, some perhaps improbable error in design, nobody was really hurt, and no real injury has been inflicted." He continued, "Well, I think, frankly, that forbidding reprocessing is a very serious mistake. First of all, all of the countries which have nuclear reactors, perhaps not all, but practically all of them do have reprocessing. So that if we don't have reprocessing, it has a negative effect. Because they exploit the reactors, and they can't put as heavy restrictions on their use as we can because we sell it cheaper, and, therefore, we can impose more stringent restrictions. Perhaps I should say next, that the only valid argument I have ever heard against nuclear energy, that's the total amount of uranium is limited. Now, this is, of course, true, if we don't have reprocessing. And, therefore, reprocessing is needed. It also makes the waste disposal very much easier if we have reprocessing." Wigner then presented various other methods for nuclear waste disposal.

Hans Bethe spoke on nuclear power development including the breeder reactor. It is powered by uranium, creating plutonium as a by-product, which in turn is a nuclear fuel. He suggested the possibility of collaborating with the French, who already have a good breeder. He indicated that "the breeder is not the only way to conserve nuclear fuel. There are advanced converter reactors. The breeder works on fast reactions. The converter reactors work on thermal neutrons, or neutrons of low energy like the light water reactor. There are many designs of advanced converters. Many of them use heavy water instead of light water."

Edward Teller made the point that it is much easier for the third world to use fossil fuels, for their initial development, than a nuclear energy-based economic development. Highly developed countries should use fewer fossil fuels and more nuclear energy. He continued by saying, "But I want to make clear, nuclear energy, as important as it is, because of the development that was ongoing ever since the second world war, and, therefore, we now have good, reliable, safe, clean, inexpensive sources of electricity." He further added, "Nuclear energy is economic only when produced in big quantities. Unless you have a plant in the neighborhood of a thousand megawatts to give an order of magnitude, to cover the needs of millions of people on American standard, nuclear energy does not turn out, in the present state, particularly advantageous."

Robert Hofstadter focused on fusion-generated electricity to come in the very near future. It was called inertial fusion process. He also stated that he subscribed to the things that have been said by the previous speakers, and to the contents of the 1977 white paper. Fusion, like solar energy and the breeder reactor, was another example of a renewable energy source. He described the inertial fusion by saying that "The inertial fusion is a process in which the heavier isotopes of hydrogen, namely deuterium and tritium are used together in a small pellet. When they react the result is a liberation of kinetic energy, as well as a stable helium nucleus and a neutron. The kinetic energy of the neutron may be used to produce heat directly, which may then be turned into electricity by conventional means, or this kinetic energy may be used to split molecules in a manner which stores energy in the form of new chemical combinations."

After the completion of all presentations I addressed Congressman Fuqua: "Mr. Chairman, this is the conclusion of our presentation to you. And we shall measure our achievement or success today in terms of the number and quality of questions we will be receiving from you and the members of your committee now." Congressman Fuqua, based on a national and international reaction to the Three Mile Island accident, asked the panel if it is possible to develop a nuclear plant that can be designed and engineered so that it is inherently safe? Teller's answer to this question was, "It seems I have heard of the word 'zero'. That number doesn't exist. Perhaps there is a zero probability that we will live forever, but, in other regards, I don't expect zero probability from nuclear plants or anything else." He continued, "Nuclear reactors are not safe, but they are incomparably safer than anything else we happen to have. From health and cleanliness alone we should choose it. And the environmentalists, if they mean what they say, should be the first ones to use it." After many questions from the members of Mr. Fuqua's committee he asked us to take a brief recess when it was already 3:30 p.m.

When we reconvened the first question was directed by Mr. Flippo to item 12 in our 1977 white paper regarding solar energy: "Is it your contention that the economic parameters of nuclear fusion are better in-hand than the economic parameters of solar energy?" My answer to the question was: "This number 12 in the white paper was essentially the creation of an expert who convinced us, that the remaining 13 of the 14 members, that solar energy is, indeed, much more expensive at present, and probably in the near-term also, than nuclear energy by a factor of four. So, therefore, expensive energy is the worst energy to live with." At this point Hans Bethe spoke. "There are many kinds of solar energy. If you talk about solar energy to make electricity on a large scale, I completely agree with Dr. Kursunoglu, it is very far in the future, and I do not see any good prospect for its becoming economically tolerable. The factor four applies to solar energy generated on earth. The solar powered satellite, I would guess the factor is a hundred." He then talked about the need for new kinds of elements to make photovoltaic cells.

Another member of the Committee, Mr. Ambro, commented to me that "You should be congratulated for bringing such a distinguished panel to this committee. However, it seems to me that one of the reasons why they are here is because of what happened at Three Mile Island." Mr. Ambro was interrupted by Mr. Fuqua "I would like to clarify that.

That has no connection. The meeting had been scheduled long before Three Mile Island." Mr. Ambro responded by saying, "I withdraw the statement." Mr. Ambro asked many questions and made many statements on his own. He asked, "Wasn't there a risk of hydrogen explosion?" To this Wigner responded, "There was no oxygen up there." Mr. Ambro: "No oxygen, so, therefore, no explosion?" Bethe intervened, "Therefore, no explosion. And one has to be very careful in talking about an explosion. First of all, I want to reemphasize there could never be a nuclear explosion."

The statements made by the Fuqua committee member Ottinger were provocative and highly anti-nuclear: "I think we have to have tremendous respect for the scientific genius that is represented at this table. But I am not sure that we can feel comfortable in relying on this panel that participated in the construction of atomic weapons that were used in Japan by the United States. We are the only country to have ever used it and caused thousands of deaths. One of the things which greatly disturbs me is the degree to which the scientific community has become a partner with the government in lies and cover-ups of the dangers involved in the nuclear field. It has just come to light that the above-ground explosions that took place as a part of our testing program in the western states had exposed the soldiers that were put there and the people in the communities there to devastating health effects."

My response was, "I participated on, August 31, 1957, in the Nevada Test Site Smokey Project. I was there after the test took place, which was a nuclear detonation 15 miles away. Two hours later I had participated in taking a trip in the test site – not on this particular one, but the one that took place a week earlier. We have seen everything. We walked over the grounds on which nuclear explosions did take place, but they were mixed up with the sand and everything. I am in very good health, Mr. Chairman. That was 22 years ago. I have no traces of anything whatsoever. Of course, I do believe my experience more than any other statement. My grandmother died at 130. She fell from an apple tree. I do not anticipate my participation in the historic event, in 1957, August 31st, is going to be my reason for my expiration. I expect not to be it, but to reach the same age as my grandmother."

I do not wish to summarize the entire course of the hearings, but I must admit that the meeting with the members of the House Committee of Science and Technology with my colleagues was an inspiring experience in learning a little bit about the *modus operandi* of an important unit of the U.S. Congress. The meeting which began at 2:00 p.m. was adjourned at 4:45 p.m. It was quite interesting that our meeting coincided with anti-nuclear demonstrations taking place outside by more than ten thousand people, as organized by the famous actress, Jane Fonda. In fact, Jane Fonda and company were received in the White House by President Jimmy Carter. We were able to visit Carter's Science Advisor, William Press but an audience with the President could not be arranged.

5. INTERNATIONAL SCIENTIFIC FORA ON ENERGY

This series of fora which began in 1977 was planned and organized, under my chairmanship, by distinguished scientists from academia and high level industrial representatives. The planning committee included several Nobel Laureates who worked hard to make successful the series of fifteen fora. The following dates, venues, topics, and financial support received provides information on the general aim and character of the fora organized under the ægis of the Center for Theoretical Studies (University of Miami) and the Global Foundation.

TABLE 1. International Scientific Fora On Energy

Date	Place	Subject
Nov. 1977	Fort Lauderdale, FL	Nuclear Energy Future of the World
Nov. 1978	Miami Beach, FL	Nuclear Energy and Alternatives
Nov. 1979	Nice, France	Energy for Developed and Developing Countries
Nov. 1980	Fort Lauderdale, FL	Geopolitics of Energy
Nov. 1981	Mexico City, Mexico	Global Energy Assessment and Outlook
Nov. 1982	Paris, France	A Global View of Energy
Nov. 1983	New York, New York	New Energy Realities
Nov. 1984	Fort Lauderdale, FL	Energy: A Non-Issue? Consequences of Being Wrong
Nov. 1985	Copenhagen, Denmark	The Silent Energy Revolution
Dec. 1986	EPCOT, Orlando, FL	Fueling the 21st Century: Today's Decisions for Tomorrow's Needs
Oct. 1987	Moscow, (former USSR)	Fueling 21st Century
Nov. 1988	Washington, D.C.	Energy Challenges: A Basis for a Global Agenda
Oct. 1989	Paris, France	Energy--Environment--Quality of Life
Oct. 1990	Miami Beach, FL	The Quest for Acceptable Energy Technologies
Oct. 1991	Washington, D.C.	Making the Market Right for the Efficient Use of Energy
Nov. 1994	Washington, D.C.	Global Energy Demand in Transition (The New Role of Electricity)
Nov. 1995	Miami Beach, FL	Economics and Politics of Energy
Nov. 1996	Miami Beach, FL	Technology for the Global Economic, Environmental Survival and Prosperity
Oct. 1997	Washington, D.C.	Environment and Nuclear Energy
Oct. 1998	Paris, France	Preparing the Ground for Renewal of Nuclear Energy

The uniqueness of the fora was the result of their scientific, independent, international, and interdisciplinary structure and the participation of people of great achievement in academia, industry, and government. The participants of the first forum in 1977 included many members of the Manhattan District project. It is also a great pleasure to reminisce about the fora held in 1978 (Miami Beach) and 1980 (Fort Lauderdale) in which Edward Teller and Eugene Wigner organized forum sessions that made a major impact in the fields that were discussed. In Edward Teller's lecture, perhaps by a slip of the tongue, he said "in the fission of U_{235} uranium atom four neutrons are produced." It was interrupted by Wigner when he murmured from his seat "two." But Teller decided to insist on "four neutrons." Wigner, politely, repeated his objection by again saying "two." At this point Teller decided to compromise by pointing at Wigner and saying "Eugene, three neutrons only, take it or leave it." The audience was elated and participated by laughing and applauding. In 1984 I invited Hans Bethe to participate in our Fort Lauderdale energy conference "Energy: A Non-Issue? Consequences of Being Wrong". He responded with the accompanying letter. Participants in these fora included many distinguished scholars: Nobel Laureates, industrialists and high level government representatives. Forum proceedings for most of the meetings have been published. In some fora white papers have been issued. Support for the above fora in the approximate amount of $1,200,000 has come from US corporations (Exxon, Phillips Petroleum, Mobil Oil, Westinghouse, Gas Research Institute, Sun Co., Electric Power Research Institute, Chase Manhattan Bank, Grumman Aerospace, Allied Chemical, Tennessee Gas, Union Carbide, Dresser Industries, General Electric, ALCOA, Cyton, Combustion Engineering, Schlumberger, Southern Bell, General Atomics, Florida Power and Light), and in the approximate amount of $120,000 from foundations such as the Energy Foundation, Alfred P. Sloan Foundation, Orleton Trust Fund, Parker Trust, and the Rosenstiel Foundation. French, Danish and Mexican governments, and the former Soviet Academy of Sciences have contributed approximately $200,000. A $20,000 grant was given by the U.S. Department of Energy. A contribution of $1,000 was made by UNESCO. I asked Hans Bethe for a report from him on the second day of the conference which would be distributed among the participants and various organizations in the US. He very kindly produced that excellent report dated February 7, 1985. I am pleased to include it here.

Floyd R. Newman
Laboratory of Nuclear Studies
Cornell University
Newman Lab. Cornell Univ.
Ithaca, New York 14853
607-256- 4397

Center For Theoretical Studies MAR 2 2 1984

March 15, 1984

Dr. Behram Kursunoglu
Center for Theoretical Studies
P.O. Box 249055
University of Miami
Coral Gables, Florida 33124

Dear Behram:

Thanks for your letter of March 1. Even though I am as busy as ever, I am tempted by your invitation to be the moderator of your session on Tuesday, November 27, on the comparison of the European and the U.S. nuclear power experience. The subject interests me very greatly.

Probably I shall not be able to come for any length of time, but I should hope to arrive some time on Monday (November 26) and leave some time on Wednesday (November 28). Rose is quite interested in coming as well.

Looking forward to seeing you at that time.

Yours sincerely,

Hans

Hans A. Bethe

HAB:vhr

CALIFORNIA INSTITUTE OF TECHNOLOGY

W. K. KELLOGG RADIATION LABORATORY 106-38

February 7, 1985

Prof. Behram N. Kursunoglu
University of Miami
Center for Theoretical Studies
P. O. Box 249055
Coral Gables, Florida 33124

Dear Behram,

Thank you for your letter of January 16 and the check for the travel expenses. Enclosed I am sending you a report of the Second Day of the Forum. I hope it is what you wanted.

As I already told you I enjoyed the day at the meeting thoroughly. I have seldom attended such a good conference. In addition, Rose and I enjoyed the evening gatherings, both on the first and second day. We had a beautiful short vacation on the beach of Fort Lauderdale. Thank you very much indeed for making this possible, and for being such excellent hosts.

Our New Year has been quite happy. I would like to wish you and your wife an equally happy 1985.

Yours sincerely,

Hans

Hans A. Bethe
Professor of Physics

HAB:ks

Enclosure

PASADENA CALIFORNIA 91125 TELEPHONE (818) 356-4272

Report on the Second Day of the International Forum.
By Hans A. Bethe.

The second day of the 1984 Forum was a great success, and conveyed alot of important information to the participants. It was devoted to a comparison of experiences of various countries in nuclear power.

The French experience has been an outstanding success. The main reporter was Pierre Zaleski. The French now get 38 Gigawatts from nuclear power. By 1990, they expect to get 55 Gigawatts which will be 75% of their entire electric supply. For comparison, the United States at present gets 14% of its electricity from nuclear power.

The cost of constructing a nuclear power plant in France was about $400 per kilowatt in 1974; it is now $800. In the U.S. presently the cost varies from $1,300 to $4,000 per kilowatt and more. The ratio of cost of building a nuclear power plant to that of coal plant is about 1.4 in France, 2.4 in the U.S. Until 1978, the U.S. and France had about the same building cost.

The net cost of producing one kilowatt hour is about 13.6 centimes in France, while coal power costs about 24 centimes. At the present time, 10 French centimes are about 1 American cent. In the United States, the cost from existing plants is about 3¢ per kilowatt hour, while a newly built nuclear plant will deliver energy at about 6¢ per kilowatt hour.

An important difference is in the use of materials which in France is only about 2/3 of the United States. Even greater is the difference of the amount of labor involved which in France is only about 1/3 of the United States. A still greater ratio exists in the number of engineers and draftsmen required. One reason for that is that a larger fraction of the design is done in the home office rather than in the field. In the United States in 1967 about 3.5 million man hours were required to build a nuclear power plant; in 1983 this number had gone up to 30 million man hours. Field services in the U.S. went from 1.3 to 22 million man hours.

The construction time in France in 1974 was about 6½ years, it has now gone down to 5 years. By contrast in the United States, it went from 5½ years to something between 7 and 14 years at present, depending on the particular plant involved.

French plants are available 72% of time from the fourth year of operation on. This availability is about the same in the United States.

What are the reasons for the excellent record of the French nuclear power industry? The main reason is that there is only one vendor, and one user. Both the construction company and the utility are stated owned. The centralization of government in France helps considerably.

Another important point is the political climate which has been favorable continually. There is agreement between all parties and among most of the population that nuclear power is necessary. There is no difficulty finding sites for placing nuclear power plants. There is national cooperation between the regulators and the utilities. There is a fixed price contract between the designer and the utility.

The presentation by Zaleski was supplemented by Vendryes. He pointed out that France is particularly favorable for nuclear power. France in 1970 imported 75% of the required energy. France has no oil and only a small amount of coal. Its water power is quite insufficient to supply a rather large country. Therefore from the very beginning, most of the French people agreed that nuclear power was essential.

It is a small enough country to have only one supplier and one user. The United States by contrast has four suppliers of nuclear power plants, many more architect engineers, and something over 50 different utilities of major size. Having one supplier and one user greatly simplifies the problem.

Because there was only one supplier, this supplier could gain a great deal of experience in designing and building nuclear power plants. They went through several families of power plants, starting with plants delivering 900 megawatts to the present design of 1400 megawatts. The key decisions are made by less than 10 people. Nuclear energy is never an issue at elections.

Howard Shapar gave a more general picture for the countries in OECD. On the average these countries get 16% of the electricity from nuclear plants. Japan now has 20 Gigawatt of nuclear plants and intends to triple this installation. In Europe generally the cost of electricity from nuclear power has an advantage compared to coal power of about 30-40%; in France it is 50%, in the United States there is at present no cost advantage of one compared to the other.

The lead times are about 5 years in Japan, similar in France, and 6 to 7 years in Canada. As already mentioned, the lead in the United States is 7 to 14 years.

Nuclear power is environmentally attractive, and this has been recognized in many countries.

For success, standardization is essential as has been shown by the experience of France and other countries.

There were a number of further contributions, including Sweden and Japan. Both countries have had considerable success with their nuclear power.

In the discussion, many people both from the United States and other countries contributed significantly. Milton Levenson of the Bechtel Power Corporation warns that standardization in the United States is not as easy as is in Europe. The reason is the very different climate in different parts of the country. You cannot design the same plant for Florida and Minnesota. Even though these remarks were well taken, it seems to me that more standardization in the United States would be a great advantage.

Levenson further pointed out that nuclear plants nowadays are often delayed not by interference by the regulators, but by lack of funds on the part of the utility.

There was an important discussion of licensing requirements related to safety. The contributors came from the U.S., from France and from West Germany. It appeared that there were some differences, but there was unfortunately not enough time to specify these differences in detail.

An additional paper was delivered at the very beginning of the session by Chauncey Starr. He pointed out that the total energy consumption in the United States has dropped from nearly 80 to 70 quad per year. The industrial consumption dropped from 33 quad in 1979 to 26 in 1983. This energy saving by industry was mainly due to the increased cost of energy.

He then showed the comparison between electricity consumption and the gross national product. If the GNP grows faster, then the electricity consumption increases by a higher percentage then the GNP. Experiences in the United Kingdom, France, other countries of OECD, Brazil, Korea and some less developed countries are similar to those in the United States.

Starr then discussed the future growth of electric capacity. He postulated that 2.9% economic growth per year is required to maintain the standard of living. On this basis, our present surplus of electric capacity will disappear by 1991, and afterwards we shall have a deficit in electric capacity. If we project 2.9% growth of the GNP, then we require by the year 2000 an electric capacity of 910 Gigawatt. On the other hand if there are no additions to electric power plants, but only retirements, our eletric power capacity will be only 565 Gigawatt. It is therefore very urgent to add new electric capacity.

I have not mentioned a number of the important contributors to the session which can be found in the original program. In conclusion, it seems to me that the future of nuclear power in Western Europe and Japan is very bright, but in the United States it is obscure.

6. MOSCOW ENERGY FORUM

The largest in the series of our energy fora was held in October 1987 in Moscow. The Moscow forum was preceded by the International Scientific Forum Planning Committee meeting, again in Moscow, in August 1986 hosted by the Soviet Academy of Sciences and sponsored by the Institute of High Temperatures. Academician Alexander Scheindlin was, then, the Director of the Institute and was most enthusiastic about holding one of our energy fora in Moscow. As the forum chairman, I presided over the planning committee meeting. It was attended by, besides the Soviet representatives, most of the members presented above. Our Soviet hosts were most attentive and hospitable. The 1987 forum program outlines were prepared during the 1986 meeting of the Planning Committee. The program focused mostly on natural gas and coal but there were also a couple of sessions on the role of nuclear energy.

Our Soviet hosts showed us films on the Chernobyl accident that occurred in April of 1986. We were impressed that the film that we watched showed how an orderly way of handling the most dangerous and terrible nuclear accident in the history of nuclear energy appeared as a routine and minor incident. We were, of course, quite impressed with the know-how of the Soviets. The film showed the containment of an accident involving the spread of radioactive clouds into neighboring countries even as far away as the northeast Black Sea area of Turkey. The graphite-moderated large nuclear-based electric power generators were known in the West as undesirable since there was a probability of a fire in the core of the reactor caused by the electron bombardment of the graphite. In fact, in Eugene Wigner's famous volumes on the theory of nuclear energy the prediction of such a release of energy was made and in the literature it was referred to as Wigner Release. Unfortunately, there are still many graphite-moderated reactors in the former Soviet satellite countries that need to be redesigned in order to make them safe for electricity generation.The 1987 Moscow meeting was attended by approximately 700 participants in which the Soviet participation constituted the lion's share of the total. There were approximately 50 U.S. participants including, besides the American members of the Planning Committee, Henry King Stanford, William Butler (University of Miami Vice-President for Student Affairs. He produced a detailed report on the Moscow Forum which included a description Soviet life in those days.), Arnold Perlmutter (Professor of Physics, University of Miami) and his wife Lynne Mayers, Ismet B. Kursunoglu, M.D. (our son) and our daughter-in-law Carole Ruegsegger, M.D, both of whom were guests of the Soviet Academy of Sciences. A part of the entourage for the 1987 Moscow forum included one of the loyal and efficient contributors to the organization, Linda F. Scott who served for 16 years as the Deputy Secretary for the CTS before my retirement from the University of Miami in 1992. The most enthusiastic forum Public Relations Director was Lisa B. Scott (no relation) who, during the forum, accompanied Henry King Stanford to the American Embassy in Moscow to secure U.S. media coverage of our forum. Lisa was also included in the forum program to deliver a short speech and to present, along with a young man from the USSR, the viewpoint of youth.

One aspect of the Russian *modus vivendi* was the exclusion of the wives from official functions; a practice also imposed on visitors. For example, during the forum I asked academician Yevgeniy Velikov, then Secretary General Mikhael Gorbachev's science advisor, if he would arrange for the members of the forum committee to be

Gorbachev's Science Advisor, Yevgeniy Velikov, is welcoming the participants to the USSR.

received by a high-level dignitary (preferably Gorbachev himself) in the Kremlin. Velikov very kindly agreed but, including my son and myself, no one could bring along their wives. However, he did, because of their official status, include the two Scotts mentioned above amongst the invitees. About 25 of us were received by the Vice-President of the USSR Academy of Sciences Yevgeniy Tarashchevitch instead of by Gorbachev. Tarashchevitch was also serving as President of the Republic of Bylorussian. While sitting on both sides of a very long T-shaped table illuminated by the beautiful sparkling glass chandeliers left over from the time of the Czars, we were served tea since Secretary General Gorbachev had already put a ban on alcoholic beverages. It was, as is customary, time for speeches. The Vice-President of the Soviet Academy said in a 10-minute talk that in the Soviet system, energy issues are as important as sustaining a Communist system.

As the Chairman of the Forum I was expected to reciprocate with equal

Conference chairman, Dr. Kursunoglu, is delivering his opening speech.

allocation of time to the Vice-President's speech. My very short comments consisted of reemphasizing the most cherished system of the West, which transcends even the energy issues, that freedom is the basis of everything. I stressed the necessity for cooperation between the East and the West on research in energy as well as among all the nations of the world and their scientists. I further stated that a "white paper" would be prepared by the Forum Planning Committee and issued to the Soviet government and the 700 forum participants on October 6, 1987.

The white paper would contain some cooperative directions and would focus on recommendations for dealing with future energy problems. After some photo sessions we left the Kremlin but none of us was aware of what was coming in the span of the next two years (i.e. by 1989 the whole of Soviet Communism would implode).

Moscow Forum White Paper

Address

to the world public by the Planning Committee and participants of the 11th International Scientific Forum on Fueling the 21st Century.

October 6, 1987
Moscow, USSR

Byelorussian President Yevgeniy Tarashchevitch is conferring in the Kremlin with the members of the forum planning committee.

The last decade of the 20th century is approaching. This century has been marked by great progress of science and technology which has changed profoundly the way of life for all people.

All of those achievements were made in parallel, and some of them directly related, with the development of energy technologies. Today, world energy consumption per year has reached an enormous level of 300 exajoules or 7 thousand million tons of oil *equivalent. The challenge today is to develop a sustainable energy system for the 21st century. How to meet this challenge was the theme of this Forum.*

In spite of the fact that all elements of the systems of production, conversion, transportation and consumption of energy are improving continuously, the growth of national incomes calls for an appropriate growth in the utilization of energy. This can be attained through an increased production of fuel and construction of new and more efficient power plants, pipelines, power transmission lines and so on.

Today, we realize that further unrestricted expansion of energy systems is impossible. This is due to the finiteness of energy resources, nonuniformity of their distribution throughout the world, high capital intensity of the fuel-and-energy complex and the ever growing adverse effect on energy on the environment.

The Planning Committee and the participants of the 11th International Forum on Energy have discussed at their sessions the present-day energy situation in the world and in individual countries, as well as advanced developments and forecasts for the future, and note the following in their appeal to the world public.

The energy needs of mankind will continue to grow further, and this is especially true of the developing countries in which the energy consumption without limiting the rate of growth of the national income in those countries. To this end, one must provide for reasonable savings of energy at all stages form the energy source to consumer.

Of special importance is the reduction of the consumption of oil and petroleum products. The fluctuation of prices in these products constitutes one of the destabilizing factors in the world economy and could lead to serious conflicts. Direct savings of oil fuel by way of improving oil-consuming devices and the replacements of this fuel with appropriate alternatives present one of the major problems to be solved in the near future.

The share of electric power in the overall energy system is increasing faster than the total energy demand. The Forum views this trend with approval, because electric power helps raise the productivity and quality of labor, introduce automation, and utilize new technologies that improve the quality of human life. Increased production of electric power will call for construction of new power plants and retrofitting of the existing plants. In both cases, one must raise the plant efficiency. An increase in the efficiency of electric power-consuming devices should be regarded as an additional source of electric energy.

The energy systems and the ecosystems are closely related. In view of the increasingly restrictive regulations concerning various harmful discharges, engineers and scientists face new problems which must be solved through the development of efficient and inexpensive technologies. The use of fossil fuel inevitably involves discharges to the atmosphere of substances among which are acid rain and carbon dioxide; the latter probably will precipitate global climatic changes. Nuclear power plants are free of some of these disadvantages. At the same time, measures must be taken to ensure reliability and safety of the operation of nuclear power plants.

The energy systems of the 21st century probably will not be a simple extension of an improved copy of what they are now. Even today one can see new trends developing; however, it is impossible to predict which ones will mature technologically and take an *important place in the energy scene. Controlled fusion, renewable sources of energy, including the transfer of energy from space, and hydrogen power may substantially change the world energy structure in the long range perspective. Looking into the future, one must give support to the development of these technologies.*

Energy development is international by its very nature. International cooperation in the sphere of energy will ensure faster progress and will make possible the realization of projects which no country could cope with on its own. Large-scale joint efforts between different nations, in particular, between East and West in the field of energy should be initiated. These efforts could include joint research on controlled fusion, and research and development on more inherently safe, as well as more reliable, fission reactors.

Energy is an important factor affecting economic and political stability. The efforts of scientists, economists, social scientists and public figures involved in the solution of current and future energy problems must be combined in order to secure the stability on the global and regional levels.

Because of the complexity of the problems related to the energy sector, it is important to approach their analysis in an interdisciplinary was for which it is desirable that the International Scientific Forum, through its Planning Committee, establish international working groups dealing with different facets such as, for example, urbanization, new energy technologies, protection of the biosphere, better utilization of nuclear energy, etc.

All of the Forum participants believe that the solution to humanity's problems, particularly the problem of energy, requires a secure and stable peace. To this end, reduction in the world's armaments could make a strong contribution, both in improving the international political climate, and in making available new resources and materials that could be applied to development of energy systems.

---- End of the White Paper ----

Finally, our impressions of the 1987 Soviet Union included the pleasure and pleasant surprise of a fleeting visit to our conference by Andre Sakharov, the great freedom fighter and distinguished theoretical physicist. Sakharov was awarded the Nobel Peace Prize but to no avail, he was not allowed by his communist repressors to leave Moscow to attend the award ceremonies in Sweden.

7. MORE MEETINGS ON THE ENERGY PROBLEM

The meetings on energy were resumed with the November 1994 meeting which took place in Washington, DC. The chosen subject was "Global Energy Demand in Transition: The New Role of Electricity". The conference was attended by about 60 participants and covered the topics described in the following Précis:

> Global Energy production, distribution, and utilization are in a state of transition toward an increased and more diversified use of electricity, which is the safest, more versatile and cleanest form of secondary energy. Electricity is easy to generate, transmit, and distribute, making its use practically universal. These facts make it urgent to explore the technological prospects and long term

> availability of environmentally benign energy sources for generating electricity. It is expected that the conference will be useful to the governments in formulating their energy policies and to the public utilities for their long term planning.
>
> The conference will: 1) assess the increase and diversification in the use of electricity; 2) assess the technological prospects for clean energy sources that still require more R & D, i.e. solar, hydrogen, nuclear (fission and fusion), etc.; 3) assess the roles of non-market factors and possible improved decision processes on energy and environmental issues; 4) make concrete recommendations regarding R & D policies and regulations to expedite the transition to a dependable, safer, and benign electricity based energy complex; 5) study the cost impact: price, environment, safety, and international security; 6) provide an analysis of an expected transition from the fossil fuel transportation to electrical transportation (e.g. electric cars); 7) the role of nuclear energy to satisfy increasing energy demand to include new technologies for waste treatment and new reactor design; 8) suggest how to optimize the use of plutonium and highly enriched uranium from dismantled warheads safely and permanently.

Some of the senior participants included Edward Teller (Lawrence Livermore Laboratory of the University of California), Richard Wilson (Harvard University), Ambassador Richard Kennedy (Washington, D.C.), Chauncey Starr (EPRI), Henry King Stanford (Former President of the Universities of Miami and Georgia), and Ambassador Gerald Clark (The Uranium Institute of London) who was the only overseas participant. The conference proceedings were published by Plenum Press, New York.

The 1995 meeting, "Economics and Politics of Energy," was held from November 27-29 in Miami Beach. Our original agenda for this meeting included:

> I. The globalization of transportation has changed the economic relationship between primary fuel resource countries and user nations. Supertankers and pipelines have made every oil field a strategic neighbor for most countries. What have gas pipelines and LNG done for natural gas' global availability and its use, arguably as an environmentally more benign fuel? Where does it leave coal? The big question is how has all this changed the interdependence of nations today with regard to assured energy sources as seen from a geopolitical viewpoint. The post cold war emergence of new oil powers in the central Asian and Caucasus Republics will make some modifications in the closely linked economics and politics of the energy situation. The newly emerging free market economies of the developing countries and their dependence on oil and hydro-energy are bound to have both political and economic impacts. The emphasis will be on trends, not on reviews of today's issues.

II. Global energy projections, technological changes such as nuclear power and the fuel geopolitics of the coming century will be the basis for political and strategic planning. Based on the scenarios of likely global economic and population growth and of new energy technologies, what are foreseeable scenarios for the geopolitics of energy a half century ahead? What fresh worldwide systems should we start now?

The political problems with profound economic impact could include, for example, the significance of the continuing worldwide growth of nuclear power, with such issues as the use of Highly Enriched Uranium (HEU) and Plutonium obtained from the dismantling of U.S. and former USSR nuclear weapons; the urgency of nonproliferation; the disposal of civilian and military nuclear waste; nuclear power alternatives.

But, the Republics of Azerbaijan, Kazakhstan, and Turkmenistan did not respond to our invitations and we had to delete the Item I (above) from the program. For the sake of illustration, I would like to include the agenda for the annual meeting of the Global Foundation's Board of Trustees and members of the Advisory Board to remind us where the world has evolved and what constitutes the fundamental issues that are of concern to everyone.

Global Foundation
Annual Meeting Of Trustees And Members
Of The Advisory Board
Tuesday, November 28, 1995
REGENCY CONFERENCE ROOM
Doral Hotel
Miami Beach, Florida
1:30 PM - 6:30 PM*

Agenda

1. A Brief Report of Activities
   (a) The status of various conference proceedings.
   (b) Dr. Kursunoglu's book, THE ASCENT OF GRAVITY (to be published).
2. New Directions in Higher Education
   (a) New curriculum to include fundamental global issues: population, environment, resources, energy, non-proliferation, war and peace.
   (b) Use of television for universities worldwide to benefit from the teaching of the outstanding scholars.
3. Future Role of Natural Gas and Electricity In Various Modes of Transportation for the Protection of the Environment
4. Nuclear Energy
   (a) Are there any new ideas (crazy enough) to lay the foundations for friendly acceptance and use of nuclear energy?
   (b) How to erase the legacy of the nuclear misrepresentations?

(c) The next energy crisis and nuclear energy.

5. Non-Proliferation
   (a) Modifications in the nuclear fuel cycle components ownership.
   (b) United Nations imposed economic sanctions.
   (c) Regional nuclear security agreements.
   (d) Non-proliferation responsibilities of the nuclear club members (US, Russia, Britain, France, and China) use of trade and economic sanctions.
   (e) Further reduction of the nuclear warheads to an optimum level.
   (f) Some new ideas.
6. National and International Terrorism
   (a) Economic and educational dimensions.
   (b) International cooperation for the prevention and control of terrorism.
   (c) Aspects of transnational migrations, economics and politics of ethnicity.
   (d) New ways of using science and technology for the prevention of terrorism.
   (e) Nuclear theft.
   (f) The role of Internet (positive and negative).
7. How to Finance Seminars or Conferences on the Above Suggested Topics

*Lunch will be served in the Madrid Room, Mezzanine, at 12 noon.

The 1996 conference focused on "Technology for Global Economic, Environmental Survival and Prosperity" to address environmentally attractive technologies for electricity production - renewables, natural gas, and nuclear energy. Recent technology developments were addressed which include creation of more efficient photovoltaic converters for electricity generation; the current and future role of natural gas in meeting global demand for electric power generation; and the status of nuclear energy, its various applications, and the prospects for its future. The Conference agenda, in light of its global economic impact, included comparative discussions of all the above alternative energy sources. The regional choice of energy sources and their impact on the global economy and environment was reviewed.

In addition to the above subjects, but strongly connected with the theme of global energy needs and security, the Conference program contained one session on new needs and directions in higher education: new curricula to cover fundamental global issues on energy, resources, and environment. Edward Teller, as shown in the letter below, was one of the Conference participants, presenting his idea of underground reactor. He also found time as mentioned in the last paragraph of his letter, to talk about the Nobel Prize in connection with protons and quarks. It is not quite clear what he had in mind, since he himself has not done any work in the field of quarks. Conference proceedings of the 1994-1998 meetings have already been published by Plenum Press.

The 1997 conference took up the most timely topic, "Environment and Nuclear Energy," and was again held in Washington, D.C. It emphasized the impact on environment of the use of energy. Could nuclear energy be a major contributor to an energy source mix that can help to alleviate greenhouse gas emissions and global warming problems? To what extent is it true that the use of nuclear energy in all regions of the world could help to solve pollution problems such as acid rain arising from heavy use of fossil fuels, particularly coal. Do the issues of nuclear waste and possible nuclear weapons proliferation present scientific, technological, and political challenges? If all these problems inhibit the global use of nuclear energy then, the world is facing an important global issue requiring a globalsolution.

HOOVER INSTITUTION
ON WAR, REVOLUTION AND PEACE

March 13, 1996

Dr. Behram N. Kursunoglu
Global Foundation, Inc.
P.O. Box 249055
Coral Gables, FL 33124-9055

Dear Behram:

I received a copy from your Global Foundation, Inc. (The document was addressed to "Board of Trustees.") and I don't know why it was sent to me.

I have one suggestion and one comment. The suggestion is either I or Lowell Wood or both of us attend your November conference in order to talk about our plan for a safe reactor. It should be underground and should be foolproof by means of excluding the fools. The point is that the reactor should be equipped to work for a few decades without refueling, without human interference and without moving parts. After the fuel is exhausted, the reactor stays underground and does not cause danger by handling its contents. I hope to hear from you as to whether such a talk would be appropriate.

My comment is about the annual workshops. I find it remarkable that missile defense has not been discussed even though it is the most obvious way to help in stability.

In an entirely different matter, I have been reading and thinking about the protons and quarks. The general way how Nobel Prizes have been given seems to be far from proper, but I obviously have no influence. The second point is that the whole discussion of quarks and gluons is a little unconvincing. I have even tried some improvements on the way to look at the problem. I have nothing really to report, but I continue to worry.

Sincerely,

Edward Teller

Edward Teller
(dictated, signed in his absence)

STANFORD UNIVERSITY • STANFORD, CALIFORNIA 94305-6010

The interest shown by Hans Bethe as expressed in his above letter, has helped to persuade others to attend and contribute papers to the conference. The conference examined the impact of nuclear energy on regional and global environmental issues under a variety of scenarios that include (1) competition in deregulated energy environments; (2) constraints levied upon use of fossil energy; and (3) possible expansion of nuclear processes into energy sectors beyond generation of electricity, e.g. process heat, fuels production, etc.

Closely coupled with the examination of environment and nuclear energy was the assessment of the overall role of nuclear energy in meeting future energy needs arising from growing world populations and economic development. Similarly, the role of nuclear power in meeting national and regional energy security objectives was addressed.

In many respects, this conference has had some historic occasions, more than the previous conferences. In particular, seeing Hans Bethe and Edward Teller, who rarely agree on anything, sitting side by side in the front row was viewed by the participants as a remarkable occurrence. Their presence in this conference greatly increased its visibility.

Cornell University
Floyd R. Newman Laboratory of Nuclear Studies

Newman Laboratory
Ithaca, NY 14853-5001

Telex: WUI 6713054
Telefax: 607-254-4552

607-255-4397

E-mail: dusty@hepth.cornell.edu

May 14, 1997

VIA FAX: 305-669-9464

Professor Behram N. Kursunoglu
Chairman of the Board
Global Foundation, Inc.
P. O. Box 249055
Coral Gables FL 33124-9055

Dear Behram:

I am interested in your October conference, because it is on a favorite subject of mine and it is in Washington, which is relatively easy to reach from Ithaca.

But I have some trouble with the program, both in general and in particular. In particular, I think Session I.A. is too heavy. You have three big guns and very little time. Moreover, none of the three of us is competent to speak on global warming; we can only speak on nuclear power.

More generally, Session I.B. is fine in idea, but needs more content. Session II likewise.

In Session III, the economics of nuclear power must be considered very early; it should not be postponed to IV.B. If Session III.A. is to discuss waste storage, it should include an expert on this subject. Professor Tom H. Pigford used to be a great expert. Panofsky does not belong here; he'd be good for Session V.B., namely international security. In Session III.B., Edward Teller on the subject of underground nuclear reactors is not a good choice. The underground reactor will never be accepted by the manufacturers or the utilities because it has higher costs and is more difficult to operate. Moreover, advocating underground reactors implies that reactors above ground are not sufficiently safe. A representative of Westinghouse would be a good choice, as is Bertram Wolfe as moderator.

Edward Teller should certainly be included in the conference; I should be happy to join with him on the panel in Session V.B. Richard Wilson is a very good choice as a moderator for that panel, or, as a participant.

EPRI is a very good source of possible participants in your conference.

I am not yet definitely accepting your invitation, but it will depend on the program and the participants.

With best wishes,

Yours sincerely,

Hans A. Bethe

HAB/dsh

P.S.: Please invite also Bernard Cohen of the University of Pittsburgh. He has done great work on the biological effect of radioactivity.

I prepared a white paper and circulated it among members of the Committee, which included Chauncey Starr, Bertram Wolfe, Edward Arthur, Richard Kennedy, Anthony Favale, Richard Wilson, Pierre Zaleski, and Glenn Seaborg. Their viewpoints and comments were taken into consideration while revising the white paper. I received a letter from Edward Teller delivered by Federal Express indicating that he would not sign it. At first I thought that this was unfortunate, since his name would add to the credentials of this white paper. I decided to revise it and, during the meeting in Washington, I asked Hans Bethe if he would consider reading and making suggestions on the paper, which he accepted. The resulting white paper was shorter than the original one and was more succinct. I sent it to Edward Teller for his comments. Of course, I did not tell him that Hans Bethe had anything to do with it. This was the strategy we decided (with Bethe) to get Teller's agreement. The next day he decided to sign up, even though he wanted some changes but I persuaded him not to change anything. I asked Glenn Seaborg to comment and accept to be one of the signatories. That worked out very well and we had, essentially, an important document. The remaining signatories accepted, after making some suggestions, to sign up also. The Nuclear Energy Institute in Washington volunteered to help us deliver 500 copies to all members of the United States Congress.

Global Foundation, Inc.

A Nonprofit Organization for Global Issues Requiring Global Solutions, and for Problems on the Frontiers of Science

Center for Theoretical Studies

P.O. Box 249055
Coral Gables, Florida 33124-8055
Phone: 305-669-9411
Fax: 305-669-9464
e-mail kursungl@netrunner.net
Website http://www.netrunner.net/~kursungl

May 28, 1997

Dr. Hans A. Bethe
Division of Nuclear Studies
Cornell University
320 Newman Lab
Ithaca, New York 14853

Dear Hans:

The enclosed program has just been updated where the two hours allocated to the first session will be partitioned with forty minutes for you, ten minutes for me, forty minutes for Chauncey Starr, and five minutes each for the annotators. The remaining ten minutes can be used for questions and answers from the floor. I hope you will find this arrangement suitable and we will be glad to consider other partitions of time that you may suggest. The program will be updated every three weeks to include various speakers with the titles of their presentations. You may complete the enclosed registration form and send it to us so that we can make the hotel arrangements for you.

I have not, as yet, spoken with Teller about his choice of topic but I am hoping that he will comply with our request not to talk about reactors placed underground. The annotators listed in Session V - A and B represent, more or less, anti-nuclear organizations. We thought to give them a chance to have their day during the discussion of the environment and nuclear energy. Any suggestions you may have to be included in the program will be most welcome.

With best regards.

Sincerely yours,

Behram N. Kursunoglu

BNK:pr

The white paper was released at a press conference on October 29, 1997, at 11:00 AM, in the Crown Plaza Hotel. There were approximately 20 participants from the press. We were competing for press attention with the visit by Chinese President Jiang Zemin and members of the Chinese delegation, who were negotiating that same day with

President Clinton to buy Nuclear Reactors from the United States. The entire press conference was organized by Laurie Cunnington, a very close friend of my family. This was the fourth time she has organized a press conference for our fora. The press conference was covered extensively in the October 30th and November 6th issues of *Nucleonics Week*. There were articles in other papers and a report on National Public Radio (NPR). Because of its importance, I would like to include the white paper at this point.

White Paper

To be released during the ***Global Foundation*** *Press Conference*
October 29, 1997 at 11:30 am in the Crowne Plaza Hotel, Washington D.C.

On the occasion of the ***Global Foundation's*** *20th Energy Conference, "Environment and Nuclear Energy", held in Washington, DC, from October 27-29, 1997, the undersigned have considered global energy needs for the future and, also, world development to meet this demand in an environmentally acceptable way.*

1. *Energy needs will increase throughout the world, particularly in developing countries due to the combination of growing populations and industrialization.*
2. *In these countries a major energy source will, of necessity, be fossil and organic fuels which will increase emissions of greenhouse gasses. This will compound already significant worldwide environmental problems.*
3. *Energy conservation in developed countries cannot adequately offset the growth in energy use by developing countries. The developed countries must, therefore, put increased emphasis on non-fossil energy technologies.*
4. *Although technological innovation may eventually provide non-polluting alternatives, at present only nuclear power is a cost effective non-fossil source of electric power.*
5. *It is therefore vital that the United States in particular, and all developed countries, emphasize nuclear power in meeting electric power needs, and to the extent possible substitute uranium for fossil fuel. It is equally critical that, as aging nuclear facilities are taken out of commission, replacement power generation be nuclear and not fossil fuel.*
6. *While we recognize the major concern attendant on widespread use of nuclear power: in particular reactor malfunction, we note that no reactor accident that harmed any member of the public has occurred in any facility meeting international safety standards (Chernobyl did not meet the standards). Fossil fuel pollution from power plants is estimated to cause 40,000 to 70,000 deaths per year in the United States alone.*
7. *The issue of nuclear weapons proliferation can be met, we believe, by strengthening the International Atomic Energy Agency both scientifically and by providing it with means of enforcement.*
8. *Technology exists to dispose of nuclear waste safely.*

9. *We hope that the Kyoto meeting will call upon all countries to cooperate in deployment of nuclear power as the available means of responsibly meeting the world's energy needs. The issues are of a global extent seeking intelligent international cooperation.*

Table 2: Signatories for the Global Foundation White Paper, October 29, 1997, Crowne Plaza Hotel, Washington, D.C.

Edward Arthur *Senior Science Adviser, Nuclear Materials and Stockpile Management, Los Alamos National Laboratory*	*Hans A. Bethe* *Nobel Laureate in Physics* *Cornell University*
Richard Kennedy *Formerly Member, US Nuclear Regulatory Commission, Ambassador At Large to IAEA, Vienna*	*Anthony J. Favale* *Director, Advanced Energy Systems* *Northrop Grumman Corporation*
William Martin *Formerly Deputy Secretary,* *US Department of Energy,* *Chairman Washington Policy & Analysis*	*Behram N. Kursunoglu* *Theoretical Physicist* *Chairman of the Board, Global Foundation, Inc.,* *Professor and Director Emeritus University of Miami*
Chauncey Starr *President Emeritus and Founder of the Electric Power Research Institute*	*Glenn T. Seaborg* *Nobel Laureate in Chemistry, University of California at Berkeley, Formerly Chairman, U.S. Atomic Energy Commission*
Edward Teller *Formerly Associate Director of Los Alamos National Laboratory, and Director Emeritus Lawrence Livermore National Laboratory University of California*	*Richard Wilson* *Mallinckrodt Professor of Physics* *Harvard University*
Bertram Wolfe *Formerly Vice President and General Manager* *General Electric Nuclear Energy*	*Pierre Zaleski* *Center for Geopolitics of Energy and Natural Resources, Université Paris Dauphine* *Formerly Nuclear Energy Attaché of the French, Embassy of Washington, DC*

This conference was described by some of the long-time participants as the best energy conference they had attended. It concluded with recommendations contained in the above white paper. The proceedings of the conference have also been published by Plenum Press for further dissemination of the ideas and trends in the field of energy among those who were not in attendance.

U.S. ENERGY POLICY AND THE NUCLEAR FUTURE

Dr. Ernest J. Moniz

Under Secretary
U.S. Department of Energy
International Energy Forum
The Global Foundation, Inc.

Thank you for inviting me to speak today. A special thanks to Behram Kursunoglu for the invitation. I would also like to acknowledge that the conference has been dedicated to the memory of Dr. Glenn Seaborg, a man who led the Department's predecessor agency — the AEC — for over a decade. Glenn did a great deal to advance nuclear energy, science and technology and we all owe him a debt of gratitude.

Energy, as the lifeblood of modern economies, is clearly a commodity that drives international considerations, be it security of oil supply or the global environmental consequences of energy use. The opportunity to share perspectives among international colleagues is important, and I appreciate the chance to update you on directions at the U.S. Department of Energy, particularly with respect to nuclear power and the electricity sector.

As is the case in many industrial economies, we are facing an especially dynamic period in the energy sector, particularly the electricity sector, as the forces of supply and demand, deregulation, and environmental protection come together.

My main focus in the energy business at DOE is in the technology arena — in finding and developing those innovative technologies we will need to meet the energy and environmental demands of the next century — and helping to define the appropriate role of the U.S government in meeting these challenges.

Our tools — policies, regulations and research and development — have *not* been adequately aligned over the last decades. I consider work on this alignment to be my principal energy-related challenge in the last year of the Administration.

Before returning to a more specific discussion about the alignment of our energy R&D with our policies and regulations, I would like to first examine three drivers that are shaping the "real world" for nuclear energy in the 21st century:

- energy supply and demand;
- worldwide privatization of energy systems, and;
- the environment.

The Challenges to Nuclear Power in the Twenty-First Century
Edited by Kursunoglu *et al.*, Kluwer Academic/Plenum Publishers, New York, 2000

GLOBAL ENERGY SUPPLY AND DEMAND

According to the DOE Energy Information Administration projections, energy demand is likely to grow dramatically, perhaps quadrupling by the end of the next century. World electricity consumption is expected to nearly double by 2020, with annual growth in the industrialized countries averaging about one and one-half percent.

Although the economic downturn in Asia that began in mid-1997 has lowered expectations for near-term growth in the region, almost half the world's increase in energy consumption is still projected to be in developing Asia.

Given the direct link between energy and economic activity, the energy sector has suffered accordingly. A range of capital intensive infrastructure projects are falling victim to the international shortfall of ready capital as projects for power generation, pipeline construction, and liquefied natural gas, for example, have been scaled back or put on hold.

To the extent that private energy investment is affected, so too are intermediate and long-term prospects for energy supply. Nevertheless, we must be confident that cyclical impacts will not affect the pressing need for greatly increased energy supply and improved economic prospects worldwide.

In the public sector, we must continue to pave the way for deployment of a broad spectrum of energy technologies — cleaner fossil fuels, nuclear, renewables. There is no "silver bullet" technology that will by itself meet our near to mid-term energy needs.

Of course, how these needs will be met will be affected strongly by other externalities or drivers, including environmental constraints and electricity deregulation.

ELECTRICITY RESTRUCTURING

Deregulation and privatization of electricity supply systems is taking place in many corners of the globe. The United Kingdom's divestiture of its energy assets was the largest privatization in history. Brazil is selling large portions of its electricity industry and expects to attract foreign investment in the $60 billion range. South Korea and Thailand are undertaking major market reforms in the electricity sector.

In the U.S., the electric power industry is currently in the midst of two kinds of restructuring — restructuring of the industry **players** in the market in response to — and in anticipation of — competition; and restructuring of the legal and regulatory rules of the game.

Newspaper headlines almost routinely announce mergers and acquisitions of electricity, natural gas and telecommunications companies, the shedding of generation assets, and new ventures in non-energy businesses — known as "convergence."

We are also, both at state and Federal levels, dismantling the legal and regulatory structures that have maintained the monopoly power of yesterday's electricity sector. Twenty four states have already implemented some form of deregulation. Industry and the states are taking the lead and moving forward, as they should. But federal legislation is still critical.

Last summer the Clinton Administration forwarded a comprehensive restructuring bill to Congress. It is our hope that comprehensive restructuring legislation will be passed in the 106th Congress which includes the major features of the Administration's bill.

Most public policy makers agree that competition in electricity supply will yield considerable benefits. We estimate annual consumer savings approaching $20 billion. But it will also likely yield benefits that are still hard to anticipate.

Remember that proponents of airline deregulation argued for price benefits of competition, but they did not foresee how it would revolutionize the logistics functions of corporate America through the rise of companies such as Federal Express. Similarly, those who advocated telecommunications competition did not anticipate the new value-added services provided at the switch and whole new categories of customer-owned equipment connected to the network.

Depending on how deregulation goes, it may introduce a significant deployment of new energy efficient technologies. The widespread use of **on-site** generation will give us an intelligent digital grid that interacts in real time with every key part of the electricity system — central station plant, transmission system, home appliances. The fact that every intelligent machine in your homes and businesses will be connected to the grid defines the scale of the possibilities.

We are optimistic that the world in 2030, while still heavily dependent on cleaner fossil fuel use, will be one in which growing demand for electricity as a preferred energy source, new inherently safe nuclear power designs, and dramatic improvements in the economics of renewable technologies and end-use efficiency, will provide a broad spectrum of clean, low-cost reliable electricity choices for the marketplace.

ENVIRONMENT

Another major externality that will shape the electricity sector is the environmental impact of energy use at all geographical scales — smog and particulates in urban environments, regional acid rain, global warming.

On a global scale, there is little doubt that human activities associated with energy production, primarily of fossil fuels, have over the last few decades, altered the composition of atmospheric gases. World carbon emissions are expected to exceed 1990 levels by 39 percent in 2010. By 2020, this figure will be closer to 70 percent. Two thirds of the total increase in carbon emissions will occur in non-industrialized countries.

Although the manner in which global warming imperatives will be implemented remains unclear, the 1997 agreement in Kyoto radically altered the nature of the debate.

Even in the absence of formal binding international implementation mechanisms for the Kyoto accord, we are seeing many businesses begin to factor greenhouse gas emission considerations into their business plans. It has simply become a good business decision and it serves customer demand. This is a profoundly important development, and nuclear power is uniquely positioned to play in this game.

The significant environmental challenges associated with increased energy use, coupled with increasing competition for fossil fuel supplies, suggest that nuclear power should continue to be an important part of the mix in planning the energy future.

In the context of these energy externalities, I would like to discuss:

- nuclear power in the competitive environment
- nuclear waste
- nuclear nonproliferation and our work with Russia
- nuclear research and development

NUCLEAR POWER IN THE COMPETITIVE ENVIRONMENT

Today, nuclear power supplies about 17 percent of the world's electricity. Ten countries meet at least 40 percent of their total electricity demand with nuclear power. In the United States, 103 nuclear power plants currently provide about 19 percent of the Nation's electric power.

In certain regions of the world, especially Asia and developing countries, where energy demand is growing rapidly, we expect growth in the use of nuclear power to continue for years to come. But worldwide, use of nuclear power will likely decrease after 2010.

No new nuclear plants have been built in the United States since 1978. In early 1998, Commonwealth Edison announced it would shut down its two Zion units; and Millstone One in Connecticut was retired last year. Utilities have shut down operating plants, and more plants are likely to be closed as the electricity system in the U.S. deregulates. The Energy Information Administration notes that, if no new plants are built, nuclear power in the U.S. is likely to drop by over 40% in the next 20 years.

Not all the news is bad however. A large number of the existing nuclear plants produce electricity at very competitive costs. Competition in the electricity sector has led to a brisk market in the sale of existing nuclear plants.

GPU sold its Three Mile Island holdings to AmerGen Energy. As already noted, Entergy recently bought Boston Edison's Pilgrim nuclear station. Both AmerGen and Entergy own and operate multiple nuclear stations, and plan to use their experience to lower operating costs at newly acquired sites.

Industry experts expect this trend to continue: single operating units being purchased by owners of multiple plants, leading to a consolidation in the U.S. nuclear industry with sustainable skilled nuclear workforces, and with the most efficient existing operations well positioned for deregulation.

In addition, nuclear utilities in the United States are also extending plant operating lives. Baltimore Gas and Electric is seeking an extension of the license for its Calvert Cliffs facility — the news out of the NRC is favorable — and Duke Power has submitted an application for the license renewal of its Oconee plant. Extension of the licenses of such economically-competitive plants can make an important contribution to meeting our environmental goals.

Consequently, the Administration has requested funding for a modest R&D program designed to help life extension and relicensing programs. The program, called the Nuclear Energy Plant Optimization program (NEPO), was developed with EPRI and will be carried out in partnership with industry.

Up front capital costs are a critical issue, particularly in our emerging deregulated electricity sector. In Japan, TEPCO reduced the construction time on its newest Advanced Boiling Water Reactor — Unit 7 of the Kashiwazaki Kaariwa Nuclear Power Station — to 51 months.

This is an impressive accomplishment, but there are demands for even faster construction. According to industry analysts, building a new nuclear plant in the U.S. would not be financially justifiable unless it could be completed in under three years. It is unclear whether existing technologies, including our Advanced Light Water Reactor designs, can currently meet this stiff challenge.

As I mentioned earlier, the climate change debate — and more broadly the debate about emissions constraints on energy production and use — is also a critical one for the future of nuclear power in the U.S.

Most of the avoided CO_2 emissions over the last 20 years have come from nuclear power. In the United States on an annual basis, nuclear power avoids greenhouse gases equivalent to burning 50,000 railroad cars full of coal.

If a true monetary value were established for carbon emissions, nuclear power could be the major beneficiary of an emissions credit trading market. Nuclear power advocates — and environmental advocates — need to play an active role in setting the regulatory framework that will advance our environmental interests.

Indeed a natural alliance of carbon free technologies — nuclear and renewables — needs to be more active in aligning energy and environment policies, such as advocating all-source regimes.

Nuclear power is clearly at a crossroads in the U.S. The challenge for the industry will be to remain viable for the next ten years, at which time, growing energy demand ... mitigating environmental impacts . . . and the need to replace aging power infrastructures . . . will create new opportunities for the industry.

NUCLEAR WASTE DISPOSAL

No issue is more critical to the future of nuclear power in the U.S. than solving the problem of waste disposal.

The Clinton Administration believes that the overriding goal of the Federal Government's high-level radioactive waste management policy should be the establishment of a permanent geologic repository — essential not only for the disposal of commercial spent fuel, but also for high-level waste and spent fuel from the cleanup of the nuclear weapons complex, and the nuclear navy.

A permanent geologic repository is also important to our non-proliferation goals: an alternative to reprocessing . . . storage for foreign research reactor fuel . . . and an option for the disposition of surplus plutonium from nuclear weapon stockpiles.

I know that there are advocates of reprocessing here today. We have concluded, however, that reprocessing is uneconomic and causes proliferation concerns. According to an Energy Resources International study, reprocessing will add about 40 percent to the price of fuel. But whether or not you share our views on reprocessing, geologic disposal of waste — for hundreds or thousands of years — is an issue that we all must face.

Last December, Secretary Richardson submitted the *Viability Assessment of a Repository at Yucca Mountain* to Congress and the President. We are on track for a 2001 suitability decision on Yucca Mountain as the location of a repository, and, assuming suitability, to submit a license application to the NRC in 2002.

And although the funding from Congress — about $60 million below our request — could jeopardize this deadline, the Secretary has directed that we make every effort to stay on schedule.

It is important to underscore that the scientific and technical work being carried out at Yucca Mountain represents cutting edge science on a first-of-a-kind project. The licensing process — for a project whose performance is to be projected over such long time scales — will also break new ground.

One of the potential ways to improve repository performance dramatically is through transmutation of long-lived isotopes. We have embarked on a roadmapping process for Accelerator Transmutation of Waste, together with colleagues from Japan, Russia, France, and other countries.

The result of this effort was an ATW roadmap released by DOE just this week. It concluded that ATW would require a six-year, $281 million R&D effort for open technical issues . ATW could complement geologic disposal, and any decision to pursue ATW would follow evaluation of technical, costs, and nonproliferation issues. Whether or not this effort leads to any ATW international collaboration, we are eager to expand international collaboration on nuclear waste issues in general.

I have just returned from an International Conference on Geologic Repositories hosted by Secretary Richardson. The joint declaration from this conference committed to continued international cooperation on waste issues and the viability of geologic repositories as one of the preferred options for disposal of nuclear waste.

NUCLEAR NONPROLIFERATION AND COLLABORATION WITH RUSSIA

I would like to focus briefly on nonproliferation issues and specifically on the United States cooperation with Russia on nuclear material security issues that are directly related to civil nuclear power issues. Our cooperation with Russia on nuclear materials issues has intensified greatly since the end of the Cold War.

A particular focus of our joint efforts in the DOE work with Minatom has been to reduce the amounts of special nuclear material and to increase the security and accountability of the material that remains. Commercial nuclear power plants are key to the program, since that is where the weapons material is ultimately "burned."

The HEU Agreement, as you may know, involves the purchase by the U.S. of the LEU extracted from 500 metric tons of HEU from Russian weapons over twenty years for use in civilian nuclear reactors. This $12 billion agreement is financed almost entirely by commercial transactions. It is probably the most significant nonproliferation action to date involving nuclear power.

On March 24, 1999, Secretary of Energy Richardson and Minatom Minster Adamov signed agreements, in conjunction with the signing of a commercial contract between Russia and three major uranium companies, that will provide for the stable long term sale of Russia's natural uranium from the original HEU agreement. This is good for nonproliferation, and good for nuclear power.

We all share a national security interest in working with Russia to assure that material removed from nuclear warheads is removed from weapons applications. Of course, there is no simple blending operation that will convert weapons plutonium into material that cannot be used for weapons without major effort.

U.S. cooperative efforts with Russia on plutonium disposition are premised on a two-track approach, including immobilization and burning as MOX in reactors. The $200 million recently appropriated by the U.S. Congress will help jump start the ongoing negotiations with Russia but, ultimately, more funding will be needed to create the necessary infrastructure in Russia to dispose of approximately 50 tons of surplus Russian plutonium, and eventually more as arms control progresses.

NUCLEAR RESEARCH AND DEVELOPMENT

To ensure a viable future for nuclear power as a key component of the world's energy mix, nuclear energy research and development is key.

As I mentioned earlier in my talk, the need to establish a greater alignment of our energy R&D investments with our policies, regulations — and the externalities I have outlined — is critical to the future of nuclear power, especially in times of declining private sector R&D.

Building on a 1997 report from the President's Council of Advisors on Science and Technology — which recommended significant increases in DOE's energy R&D investments — we undertook a detailed examination of our energy R&D portfolio to see precisely where we were investing our research dollars. This process forced us to take off our program-specific blinders and, instead, align our R&D activities against high-level strategic goals, such as "*ensuring against energy supply disruptions,*" or "*improving the efficiency of energy systems.*"

The result, the *Energy Resources R&D Portfolio* —or "boxology" inside DOE —has shown us where our investments are currently going—and where things might need to change.

Following through on this effort, the Department commissioned a panel of some of our most experienced laboratory and senior technical officials, aided by academia, to analyze our energy R&D portfolio by asking and answering the question: *Is the portfolio likely to produce results that are needed to make signifcant progress towards achieving our strategic goals?*

This analysis identified several significant gaps and opportunities in our R&D investments, the most relevant to today's discussion being a gap in "*Maintaining a viable nuclear energy option.*"

To fill this gap, we will need to increase our nuclear energy R&D to cover the complete spectrum of research needs . . . from power generation . . . to non-proliferation . . . to waste disposal. The Department's Nuclear Energy Research Advisory Committee — NERAC — is currently working on an analysis of nuclear R&D needs. We hope that this effort will further inform and focus our nuclear energy R&D needs and help us fill our portfolio gaps.

We are already making progress. Many of you are aware that the Department of Energy introduced a new initiative last year — the Nuclear Energy Research Initiative, or NERI. In the first year of the NERI program, the Department awarded forty-six research grants to different organizations, including universities, national laboratories, and industry organizations.

Three of these awards were for proliferation resistant reactors. By exploring advanced concepts such as modular reactors with long-life cores and thorium-based fuel cycles, we may be able to find solutions to the greatest challenges facing the nuclear energy industry.

International collaborations will also add tremendous value to our research and development efforts. Indeed, another activity in progress right now is a look (by NERAC) at our future infrastructure needs to support the domestic and international nuclear establishment. We would be happy to discuss our findings in an international context.

While the NERI program was only funded at $19 million in FY 99, we received over $300 million in proposals for the first round of awards — there is obviously significant need and demand for nuclear R&D. Ultimately, we hope that the research developed under the NERI program, and the work of NERAC, will help us to achieve our goal of a secure energy future, with nuclear as a competitive player in our energy mix.

CONCLUSION

The news then on the future of nuclear power in today's energy environment is mixed. Good, because nuclear power will continue to be an option, particularly in Asia . . . because construction costs and times are coming down . . . because some units are able to compete in a restructured market . . . because as a non-emitter of carbon, nuclear power is more desirable in a carbon-constrained environment.

Significant problems do exist however. We need only to look at two issues we have been dealing with at DOE — one international and one specific to our complex — to understand the major difficulties nuclear energy faces in the years ahead.

The accident at Tokaimura, Japan, and accompanying headlines such as "*Can it Happen Here*" or "*Where is the next Chornobyl*" illustrate that the public's fear of nuclear power remains and that an accident **anywhere**, affects the industry **everywhere**.

The DOE technical team that went to Japan after the accident concluded that, in spite of a relatively strict regulatory regime, human error — a lack of training and adherence to procedure — was the major cause of the accident. The industry must redouble its efforts to train workers and make certain that the rules and regulations are followed.

The difficulties the Department of Energy is having in deciding what to do with its Fast Flux Test Facility provides another example of problems the industry faces.

The Department has expressed no preference on whether to restart this facility and is going through an EIS process to determine what value, if any, restart would bring to our nuclear R&D infrastructure.

The decision to simply seek an answer to this question has caused a significant amount of controversy in the region and reflects the deep ambivalence many in the public continue to have about nuclear research specifically, and nuclear power in general.

I believe that nuclear energy must be part of a comprehensive integrated discussion about addressing our national and multinational goals in energy supply and security, environmental stewardship, and economic development.

We must redouble our efforts to address these issues and concerns and commit to investing in the technological solutions that will be required to ensure the future of the industry. I look forward to working with you to do so. Thank you.

CHALLENGE TO NUCLEAR POWER IN THE NEXT CENTURY CAN THEY BE OVERCOME?

Peter Beck, Malcolm Grimston

The Royal Institute of International Affairs
10, St. James's Square
London SW1Y 4LE

INTRODUCTION

A paper given to the Global Foundation Conference in Paris last year[i] came to the conclusion that

a) In the light of the many uncertainties of the future, there is a strong case for keeping the nuclear option open

b) This would require a far better understanding between the public and the industry.

c) To achieve this would require a look at what an acceptable nuclear industry might be like and whether the necessary technology for this can be developed.

What the talk did not address was whether and how aims b) and c) might be achieved. It is the purpose of this paper to remedy this omission. It will, first, analyse recent developments affecting energy in the political and economic fields. That is to be followed by considering the obstacles to the achievement of greater public support and a discussion whether these can be overcome. Lastly, we draw some conclusions

RECENT DEVELOPMENTS IN THE ENERGY FIELD.

There have, over the last decade, been at least three major developments in the energy field, all with considerable potential effect on the future of nuclear power:

a) The debate about climate change, leading to acceptance by many (but by no means all) countries of the need to reduce the emission of greenhouse gases, especially CO_2.

The Challenges to Nuclear Power in the Twenty-First Century
Edited by Kursunoglu *et al.*, Kluwer Academic/Plenum Publishers, New York, 2000

b) The pressure to deregulate energy markets, especially electricity and so changing this market from being monopolistic and highly controlled to one fully competitive and operating with minimum government interference.

c) The development of gas-fired power stations, which, by making use of gas turbines and a combined cycle can achieve far higher efficiencies than normal power stations using steam turbines. In addition, world-wide reserves of natural gas have doubled since 1980.

Of these, climate change has the potential to become by far the most important. If taken seriously by politicians (which is not the case in a number of countries), this could require eventual stabilisation of greenhouse gas concentration in the atmosphere (usually expressed in terms of ppm CO_2 equivalent) if the environmental impact is to be controlled. Should such a stabilisation target be set at double the CO_2 content at the start of the industrial revolution and this to be reached by 2050, calculations show that fossil fuel may by then have to be limited to around 25% of energy demand, compared to some 82% today.

By 2050 total energy demand may have increased by a factor of 2 or 3 from today's. So, in order to meet such a condition, carbon-free energy supply would have to have grown by a factor of nearly 15. In that time-span the only sources of carbon free energy are renewables, such as solar, biomass and wind, sequestration (exclusion of CO_2 from the atmosphere) and nuclear fission. There is considerable debate, but no conclusion, how far and how fast renewables might grow, but clearly, if economic and acceptable, here is a major potential opening for nuclear power.

Should for one reason or another the climate change issue disappear, there would only be a future for nuclear power if it can compete vis-à-vis combined cycle natural gas plants. At least for the next 50 years, there should be adequate gas resources and if gas hydrate deposits are taken into account, resources could last centuries.

The second development, the change within the electricity generating industry, from being a regulated monopoly to one operating in a free market, is having a profound effect on the structure of the electricity industry and therefore on nuclear power, especially where decisions about new capacity are involved. In a regulated and controlled utility market profitability may be low, but it is secure; there is a captive market and within reason over-expenditure of capital or minor inefficiencies can be passed on to the customer. That is not the case in a competitive market. In such a market the effect of cost- overruns, delays, inefficiencies, low plant availability cannot be passed on to customers and will, therefore, affect profit and hence the shareholder. In such a market the best producer may well be highly profitable, but even the average producer could well be struggling.

Under such circumstances, companies tend to choose relatively small generating units with low capital investment per unit of capacity, which are quick to build and have a short pay-out. These choices reduce the overall risk of an investment and are today available by choosing combined cycle gas fired plants if natural gas can be obtained at competitive prices. Compared to present nuclear power plants, such units have 1/3 of capital/kw, take two, rather than four to six years to build. Being more acceptable to the public, suitable sites for such plants are also far easier to find.

Such units are, of course, vulnerable to fluctuation in the price of gas. However, the gas market is moving in the direction of there being a recognised global price for gas (as presently for oil), from which prices in specific areas can be derived. It can thus be argued that as long as gas remains a major fuel for power production, significant changes in gas price will affect most competitors and can be passed on to customers, Such risks need, therefore, not be borne by the power producer and its shareholders.

OBSTACLES TO THE USE OF NUCLEAR ENERGY.

The popularity of nuclear power is low in many countries; four main factors tend to be advanced as the reason for this:

a) Fear of radioactive release and of more general dangers posed by nuclear plants
b) Unattractive economics compared to other energy forms
c) Concerns about the risks of weapons proliferation
d) Lack of agreement about the destination of long-lived nuclear waste.

As regards *radioactivity and safety*, the industry sees the dangers of radioactivity and of plant safety as greatly exaggerated; it believes that, as a result, it is over-regulated. Disagreement is concentrated on two fronts. Firstly, on plant safety, the opposition, whilst perhaps accepting the industry's strong commitments to safe plants, believes that one will never be able to guarantee the high quality of plant management, operation and maintenance necessary to ensure safe operation and they can point to many examples of such failures. In any case, never having had an accident cannot prove that there will never be one.

When it comes to radioactivity, there is scientific controversy about the effect of low levels of radiation, i.e. levels well below those of natural radiation. Some believe the effect to be zero or even positive, whilst others suggest that any increase in radioactivity, however small, increases the chance of cancer.

The fundamental problem is that statistical methods, only, are inadequate tools in trying to determine whether a small increase in radiation affects cancer rates. Any effect is likely to be lost against the natural variation in the disease. Once more information becomes available about the mechanism by which radioactivity causes cancerous growth, it may become easier to resolve this issue, but there is little indication when this might happen.

In the meanwhile, the precautionary assumption continues to be used by regulators that there is a linear relationship between the level of radioactivity and the risk of cancer even at very low levels. That has a considerable effect on plant design and costs, but it is difficult to see how this could be changed whilst there is such disagreement within the scientific community[ii].

Regarding *economics*, it was already mentioned earlier, that the competitive position of new nuclear power plants has deteriorated over the last decade owing to the development of gas-fired combined cycle generating plants and the effect of deregulation of the electricity market. The generating industry is looking for reactor designs of lower cost per unit of capacity, higher efficiency and flexibility regarding scale, but most developments on offer for the LWR are for large scale plant which are unlikely to achieve this.

The nuclear industry argues that present economic comparisons are flawed because there is no 'level playing field'; the cost of nuclear power has to include the cost of waste disposal and of decommissioning, but gets no offsetting bonus for not causing emission of greenhouse gases and other gaseous pollutants. Whilst that is true and may well change over the next decade, there must be doubts whether that alone could radically change the situation. High capital investment, large scale and lengthy building time considerably increases the risk of an investment in a competitive market even should the playing field be beautifully leveled.

The danger of weapons *proliferation* owing to the spread of nuclear energy was recognised early on and was one of the main reasons for the establishment of the International Atomic Energy Agency with its role of setting security standards and of monitoring.

However, with the experience of Iraq and N. Korea in mind, there is a belief in anti-nuclear quarters that, should a state wish to develop nuclear weapons, access to civil nuclear technology can be a clear advantage. India and Pakistan appear to be additional examples of that.

In addition, there is the fear of rogue states or terrorists spiriting away plutonium containing materials that, when not highly radioactive, could be a ready source for the few kgs of Plutonium needed for a nuclear weapon. After some decades of cooling off, spent fuel could be such a source and so could fresh MOX fuel. This danger does not seem to be treated as seriously in Europe as in USA, where, during the 70s, it was one of the main reasons for abandoning reprocessing of spent fuel and the development of the fast breeder reactor. Whether such dangers are realistic or not, they do provide a discernible reason for concerns about the continuation and possible expansion of nuclear energy.

Disposal of *long-lived nuclear waste* seems to be one of the major concerns about the industry in the mind of much of the public. The industry's arguments that such waste can be dealt with safely and securely for 100,000+ years (a time-span never previously mentioned about any other industrial issue) by deep burial in stable geological formations may well have the opposite effect to that intended. It can lead to sceptics saying and being believed that such waste must be very dangerous indeed, but as it is impossible to look such a long time-span ahead, the projected solution may well be untrustworthy. The argument by anti-nuclear lobbies that such repositories would become the potential plutonium mines of future generations, an argument difficult to refute, exacerbates the concerns and leads to additional worries about proliferation.

Comparison with other dangerous waste, such as from the chemical industry, appears to have little effect, possibly because that issue has been with us for a long time and is rarely in the headlines.

In summary, the negative perception about all four issues seems strong and is not easy to refute; the differences between the pro- and anti groups are in differences of perception, not provable facts. As a result, there have to be doubts whether development, say, of safer designs of reactors or repositories will greatly affect public perception. What can?

CAN PUBLIC SUPPORT BE REGAINED?

Perhaps the first prerequisite to gain public support is that there should be demand for nuclear plant by the generating industry. Should that not materialise, public support becomes irrelevant; the industry will fade away. When, in the 1950s and 60s civil nuclear power was first developed, it was seen as the means of producing cheap power in a way which could guarantee security of supply to a country. It also represented new technology at a time when new technology was worshiped. Admittedly, its connection with the atomic bomb caused early resistance, but that was insufficient to sway the enthusiasm of governments and industry. At the time, therefore, the generating industry was keen to enter this field and the public could readily see the benefit of nuclear power; only a few warned against its dangers.

Today's situation is virtually the reverse. No new nuclear power plants are under construction in countries that have a competitive electricity market. Also, because of World Bank and other lenders' reluctance to assist construction of nuclear plant, there are questions how many of the 25 or so reactors, now under construction, will be completed. In countries where public opinion matters, people perceive the risks, but see few benefits, whilst the electricity industry and governments, with a few exceptions, such as France and S. Korea, are too concerned about the vociferous opposition to this power source to do anything, but 'sit on the fence'. Yet, another necessary, though by itself an insufficient requirement for changing the public attitude, is strong government support for nuclear power.

Most OECD countries have little need for new generating capacity at least until the present nuclear plants reach the end of their life – say from 2010 onwards. Then choices will have to be made between nuclear, renewables and fossil fuel plants, but if nuclear is to have a chance, the industry will have to be able to offer designs that suit the needs of the electricity industry. Bearing in mind the effect of deregulation, there is quite a possibility that this may no longer be the large base-load unit of 1 to 2 Gw(e), but a more flexible modular design which could compete with combined cycle gas plants. Design studies of LWRs have so far concentrated on large-size units, although some work is being done on 600Mw(e) size. Perhaps it is now time to re-think priorities in this area.

There are other alternatives. Schemes are well advanced to build 100Mw(e) High Temperature Gas Cooled reactors in S. Africa which make use of pebble bed technology, helium for cooling and closed cycle gas turbines for power generation. If present estimates are confirmed, the capital cost/kw of such a reactor could be about half that of a large scale PWR, building time perhaps three years and efficiency around 50%. If this design proves itself, it could well have a large market in the developing world and in areas where natural gas is either unavailable or expensive and might even become competitive with natural gas combined cycle plants in OECD countries.

Further away in time are possibilities of using fast reactors, though, at least for some decades, not as breeders. The Soviet navy has been using such reactors, using a lead/bismuth eutectic mixture as coolant, for some decades in some of their high performance submarines and it is understood that work is now going on to see whether this design could be made suitable for small commercial power production.

There are also technological means under study that may reduce the risk of proliferation by making changes to the fuel cycle. One such scheme makes use of partitioning and transmutation with the aim of destroying all the plutonium and other minor actinides as well as producing waste for long term storage which should need secure isolation for some 300 years instead of many millennia if spent fuel is stored. The waste would also not contain any material suitable for proliferation. Such changes should make such storage a little more palatable[iii].

The problem with all such ideas is that they are only ideas. Before they can be turned into real possibilities, considerable R. & D. would have to be undertaken, perhaps lasting 10+ years. If successful, which of course cannot be guaranteed, commercially sized demonstration units would have to be built and operated satisfactorily before the generating industry would be willing to take the risk of investing in these developments. All this implies that with the possible exception of the modular HTGR, it may be 20.years or more before the first commercial plants come into operation. We may well have the time, but ways will have to be found to find the money.

In the past, such funds tended to come from governments, but under the concept of deregulation and leaving major decisions to market forces, governments appear to believe that research will, in future, be funded by the market. However, no commercial company operating in a competitive market can afford to spend large sums on research that, even if successful, will not provide a return for some decades. It is of note that the US Dept. of Energy has recognised this and has started a new R.&D. initiative, NERI (Nuclear Energy Research Initiative) to support just such innovative research as mentioned in this section. The Department is also looking for widening international collaboration with companies and/or governments in this area.

In this connection, a recent report on the future of nuclear energy by the UK Royal Society and the Royal Academy of Engineers[ii] came to the conclusion that the issue of climate change will require vastly more research to find how best to develop and make use of non-carbon energy sources than presently planned. The Report suggests that the world needs a mechanism for international collaboration to ensure that adequate funds for such work, which would cover renewables as well as nuclear energy, become available and that they are spent effectively. They believe that such an effort will need strong partnership between governments and industry.

CONCLUSIONS

- The analysis in this paper comes to the conclusion that changing the public suspicion of nuclear energy is well-nigh impossible unless and until the electric generating industry perceives that the nuclear industry has a product which it requires. As a result of the push for deregulation this may no longer be the large base load unit, but a more flexible, smaller and cheaper plant.

- The question of waste storage and proliferation will also have to be clarified in ways that reduce the present concerns.

- Another requirement is that governments must be willing to back the power generators in their choice. As a minimum, they will have to provide a framework for penalising emission of greenhouse gases and establish a regulatory safety regime which brings all sources of energy into line.

- There are technological possibilities for achieving these goals, but they will require considerable R&D and commercial demonstration before they become real. This will demand time - perhaps 20 or more years - and a level of funds that to a large extent will have to come from governments.

- Because of the high demand for research into all non-carbon energy sources, international collaboration may well become essential.

- Once all these requirements are met, it may be possible to get to a point when the public will see that the advantages of nuclear power outweigh the risks. Only then might it reduce its opposition.

- It is unlikely to do so, if the industry stick to its stance (whether correct or not) that the present technology is the best, that if the playing field were level, it would be competitive and that past records show it to be at least as safe as competitive energy sources.

---

[i] Peter Beck, *The Global Energy Situation in the Next Century*, Preparing the Ground for Renewal of Nuclear Power, Ed. Kursunoglu et al.,Kluwer Academic/Plenum Publishers, New York, 1999
[ii] Joint Working Group, *Nuclear Energy, the future climate*, Section 10.3, The Royal Society and The Royal Academy of Engineers, London 1999
[iii] F. Venneri, '*Disposition of Nuclear Waste using Subcritical Accelerator--driven Systems: Technological choices and implementation scenario*' Proceedings of the conference on new approaches to the nuclear fuel cycles and related disposal systems. International Science & Technology Centre, Sarov, Russia, June 1998.

NUCLEAR TECHNOLOGY: NEED FOR NEW VISION

Clinton Bastin

Chemical Engineer - Nuclear Programs
United States Department of Energy (Retired)
Signer and Director of the Eagle Alliance[1]
987 Viscount Court
Avondale Estates, GA 30002

Nuclear power plants generate electricity without release of chemicals that cause atmospheric pollution and global warming that threatens catastrophic climatic changes[1]. Increased and more efficient use of nuclear technology could reduce this danger and threat. However, in its recent report[2], Mobil Corporation dismisses nuclear power as a non-fossil alternative for reducing emissions that cause global warming, because of "concern over safety and proliferation."

Mobil's vision for nuclear power is flawed. But it is based on a 1998 International Energy Agency report[3], reflects the vision of many Americans and their political leaders, and is reflected in US nuclear policies and programs. The vision is flawed because:

- Well-funded organizations provide misleading information which overstates dangers of nuclear technology to local public interest groups, news media, local, state, and national political representatives and others. This information is provided under the guise of bringing "scientific excellence to public policy issues to promote the democratization of science and a healthier environment."[4]
- The nuclear community has failed to provide full and accurate information to these same entities. In particular, the nuclear community has failed to provide information to the public about the highly successful, coordinated, worldwide efforts by nuclear utility organizations over the past ten years to improve safety and performance of nuclear power plants.

---

1 The Eagle Alliance is a partnership of individuals and representatives of corporations, universities, unions and other organizations who have worked to develop peaceful uses of nuclear technology and believe that nuclear technology is a proper, safe and essential element of advanced civilizations. The mission of the Eagle Alliance is to provide full and accurate information to Americans about the great benefits of nuclear technology, and to correct misinformation.

The vision is also flawed because the nuclear community has failed to make clear to the public and political leaders:

- Differences between actions that led to nuclear proliferation and those for nuclear power, and the lack of a credible proliferation threat from well managed and well safeguarded nuclear power.
- The fact that spent fuel reprocessing and recycle are essential components of good nuclear non-proliferation and radioactive waste management practices. These actions are needed so that more efficient use can be made of fissionable materials, and unwanted radioactive fission products can be disposed of without need for permanent safeguards. In addition, potential weapons usable materials are destroyed through beneficial use.
- The fact that fast reactors are an essential component of good nuclear non-proliferation practice since they destroy through beneficial use, nuclear material and source materials that could be used to make weapons.

A new vision is needed, based on accurate assessment of experiences with nuclear technology, lessons learned from those experiences, efforts to achieve outstanding safety and performance, and good policies and management. This new vision - and communication of that vision to Americans and their political leaders - will lead to recognition of the need for the great benefits of nuclear technology, and resumption, by the United States, of its appropriate role as a leader in its use.

This "new vision" will differ from the present vision - but is recaptured from the exceptional vision of past leaders of nuclear programs - including Glenn Seaborg, to whom this conference is dedicated. However, it rejects the vision of those that led to failure.

THE NEW VISION INCLUDES FULL COOPERATION AMONG NATIONS FOR ALL COMPONENTS FOR PEACEFUL USES OF NUCLEAR TECHNOLOGY.

Early leaders of the United States Atomic Energy Commission recognized that uncontrolled development of nuclear technology by individual nations could lead to nuclear weapons capability in many nations, and insecurity among virtually all nations about the nuclear intentions of their neighbors. Accordingly, they proposed to President Eisenhower and The Congress a program of "Atoms for Peace." Their vision was that the US should share nuclear materials and technology with other nations for peaceful uses so that an international safeguards regime could be developed to provide assurances that nuclear materials were not diverted to weapons programs. The Congress enacted and President Eisenhower signed the Atomic Energy Act of 1954, which provided for "Atoms for Peace."

Their vision for the program as a basis for international safeguards and safe, cost effective, peaceful uses of nuclear technology was sound; vision for its implementation was not. The first "Atoms for Peace" were sixty tons of heavy water supplied by the US to India for use as a moderator in the Canadian-supplied "Cirus" nuclear reactor. Cirus was called a research reactor, but in fact it was a weapons material production reactor. It was modeled after Canada's "NRX" reactor which was used for many years for production of plutonium for US nuclear weapons under a mutual security agreement. Later, the US supplied low-cost, Oak Ridge National Laboratory (ORNL) "pilot-plant" reprocessing technology to India and other nations, and a reprocessing pilot plant design to India.

In 1974, India detonated a nuclear explosive made with plutonium produced in the US and Canadian-supplied facilities, which were operated by India outside of the international safeguards regime. Leaders of India claimed that the device was a "Peaceful Nuclear Explosive," similar to those tested by the US under its "Plowshare" program, but other nations gave little credibility to this claim. Despite this major affront to the international non-proliferation regime, Presidents Nixon, Ford and Carter; leaders of the USAEC, Department of State and other national security agencies; and the Joint Committee on Atomic Energy of the US Congress continued to recognize the value of international collaboration in appropriate, peaceful uses of nuclear technology as a basis and an incentive for enhanced international safeguards. They also recognized that operation of the two US-supplied light water reactors in India under international safeguards - including well-safeguarded reprocessing and recycle of LWR spent fuel - was not a proliferation threat. Accordingly, they continued to support the supply of nuclear materials for operation of nuclear power plants in India, and maintained high priority bilateral and international efforts for enhanced safeguards in India.

At the end of 1974, programs of the AEC were transferred to the Nuclear Regulatory Commission (NRC) and the Energy Research and Development Administration (ERDA, later the Department of Energy, DOE), and responsibilities of the JCAE were assigned to other Congressional Committees. Leaders of the new Federal agencies did not recognize the value of continued collaboration and efforts for enhanced safeguards, and supported legislation in Congress for the Nuclear Non-Proliferation Act (NNPA), which precluded US nuclear assistance to nations not a party to the nuclear Non-Proliferation Treaty (NPT). The NNPA also precluded assistance to other nations in technology essential to best non-proliferation practice and good management of spent fuel from commercial nuclear power plants.

As one of his final acts, President Carter took advantage of an exclusionary clause of the NNPA, overturned denial by the NRC of an export license for low enriched uranium for India, was narrowly upheld by one of two houses of The Congress, and was soundly criticized for his lack of support for nuclear nonproliferation. Subsequent US Presidents have been advised to continue to support the isolationist provisions of the NNPA, at great cost to development of technology for more viable nuclear power, and to global security. Quality leadership by the United States for collaboration with India for peaceful uses of nuclear technology in parallel with efforts for enhanced safeguards could well have precluded the recent nuclear tests and nuclear saber rattling of India and Pakistan[5].

The Institute of Nuclear Power Operations (INPO) was created in 1979 by the US nuclear power industry to promote the highest levels of safety and reliability - to promote excellence - in the operation of all US nuclear power plants. A companion organization, the World Association of Nuclear Operators (WANO), formed in 1989, unites every commercial nuclear power plant in the world for similar goals for excellence, with members from all 32 countries that currently operate some 430 reactors. These efforts have been highly successful in improving safety and productivity of nuclear power plants throughout the world. INPO and WANO are an excellent model for "Atoms for Peace" for all systems needed for viable nuclear power, and other peaceful uses of nuclear technology.

THE NEW VISION INCLUDES ASSURANCES THAT BEST TECHNOLOGY WILL BE AVAILABLE FOR PEACEFUL USES OF TECHNOLOGY BY ALL NATIONS, AND THAT FULL INFORMATION ON SUCCESSFUL AND UNSUCCESSFUL EXPERIENCES WILL BE PROVIDED.

At the beginning of commercial nuclear power, nuclear power plant operators and vendors, and leaders of the AEC, recognized the need for appropriate management of spent fuel, including reprocessing. Reprocessing was needed so that fissionable materials could be recycled into existing and later advanced nuclear power plants and destroyed through beneficial use. Unwanted, intensely radioactive fission products could be isolated indefinitely from the biosphere in well-engineered repositories without need for permanent safeguards - which would not be possible. Nuclear power plants were designed with spent fuel storage capabilities that reflected the need for reprocessing and recycle of spent fuel.

In 1957, the AEC adopted a policy for receipt of spent fuel from commercial nuclear power plants - including those in other nations supplied by the US. Some AEC leaders recognized the difficulties of safe and successful reprocessing, and the nuclear proliferation threat of wide-scale deployment of low-cost, pilot plant reprocessing by nations with limited nuclear programs. Accordingly, responsibility for receipt of spent fuel was assigned to the AEC's most successful reprocessing site, the Savannah River Plant (SRP) in South Carolina.

Facilities were built for receipt and storage of spent fuel, clearance was obtained from major ports for its import, negotiations were carried out with nuclear utility operators and suppliers for its acceptance by the AEC, and research and development was carried out to permit reprocessing in H-canyon, the most successful and versatile reprocessing plant at SRP.

Terms for settlement were based not on actual AEC reprocessing costs, but estimates by ORNL for a conceptual reprocessing plant[6] with a capacity of one ton per day, operated 300 days per year, i.e., 80% time operating efficiency (TOE). Capitol cost for the conceptual plant was about $20 million; unit reprocessing cost was about $20 per kilogram of uranium.

The conceptual design and operation were stated to be "based on experiences AEC has accumulated in its operations."[7] In fact the design was based on the ORNL pilot plant concept incorporated in the Idaho Chemical Processing Plant which cost about $20 million; TOE was based on experiences at Hanford and SRP in plants that cost about $100 million. The TOE for the ICPP was not 80%, but 3%[8], and there were major safety deficiencies.

AEC vision expressed in its policy statement was that reprocessing of power reactor spent fuels would be done by the commercial sector, at costs approximating those of the AEC-ORNL conceptual plant. The "Industrial Reprocessing Group" studying reprocessing accepted the AEC vision, which led to the ill-fated Nuclear Fuel Services (NFS) reprocessing plant at West Valley, NY. Wes Lewis, who had managed the ORNL pilot reprocessing plant, was, under the circumstances, an outstanding manager for West Valley. A TOE of almost 30% was achieved, but process losses and radiation exposures to workers were more than a factor of ten larger than those at SRP and final product often failed to meet specifications. During the sixth and final year of operation, radiation exposures were well above requirements and rising exponentially, release of radioactivity to surface streams exceeded technical specifications, and AEC regulatory authorities ordered a halt of operations[9].

General Electric Company designed and built the Midwest Fuel Recovery Plant (MFRP) at Morris, Ill., and planned to start operation in 1972. However, technical problems during cold testing led to a corporate review, with conclusions that the combination of more complex processing equipment with higher expected failure rates, and close-coupling of process steps, which required much longer time to resume operation after shutdown, would permit a TOE of only a few percent. GE decided not to operate the facility[10].

In 1967, Allied Chemical Company (ACC) assumed responsibility for operation of the ICPP. Corporate officials read company reports of ICPP operations which indicated economically attractive reprocessing. General Atomics Corporation (GAC) was attempting to commercialize High Temperature Gas-cooled Reactors (HTGRs) which required reprocessing for viable operation; favorable fuel cycle economics were based on reprocessing in a conceptual plant designed by ICPP staff. In 1970, ACC and GAC formed a subsidiary organization, Allied-General Nuclear Services (AGNS) to build and operate a commercial fuel reprocessing plant near Barnwell, SC, the Barnwell Nuclear Fuel Plant (BNFP), using the ICPP as a model. Arnold Ayers, production superintendent at ICPP, was selected as technical manager for BNFP.

During the period 1972 to 1974, AEC reprocessing staff told ACC executives that safety of operations at the ICPP was unsatisfactory and productivity was overstated by a factor of five, and made arrangements for assistance from another AEC reprocessing contractor to help resolve major problems. This staff also suggested to GAC staff that cost estimated by ICPP for an HTGR reprocessing facility was under-estimated by about a factor of ten. GAC contracted with Bechtel Corporation for an HTGR fuel reprocessing plant design study, which confirmed the higher costs. After an expenditure of $500 million, GAC withdrew from HTGR commercialization. The parent companies were then fully aware of deficiencies inherent in the model they had selected for their reprocessing plant design, and AGNS informally notified the AEC that it would not complete the BNFP without substantial Federal support.

In 1974, after failure and indications of failure of three commercial reprocessing ventures, the AEC reassigned programs for support of commercial fuel reprocessing to emphasize successful experience and lessons learned from that experience. Responsibilities were transferred from the AEC Division of Reactor Development and Oak Ridge National Laboratory with their pilot plant reprocessing model, to the Division of Production and DuPont Company-operated SRP with their safe, successful production-scale reprocessing experience.

DuPont carried out and supported research and development by others focused on conceptual design studies for an NRC licensed fuel recycle complex based on its successful reprocessing experience and lessons learned from that experience and the experience of others. The design studies were completed and reports issued in November 1978. Costs for the 3000 ton/year integrated fuel reprocessing/fabrication facility were estimated at $3.7 billion (1978 dollars). Special features of this facility design include:

- Canyon structure for containment of process equipment, which would be installed and replaced remotely by overhead cranes. This arrangement is most cost effective in that it provides for maximum use of building space since there is no need for space between process equipment. Failed equipment can be replaced in less than one day and is then moved to separate maintenance shops within the canyon structure for decontamination and repair.
- Use of best technology, including centrifugal contactors for first cycle solvent extraction and solution storage between process steps. This permits process operation at full capacity within a few minutes after startup, compared to eight days at Hanford PUREX and thirty days for the ICPP, and is a major factor in achieving the high time operating efficiency of 80%. This also permits efficient operation at reduced capacities, thus avoiding accumulations of accessible weapons usable materials.
- Product recoveries of greater than 99.8%.
- Spent fuel reprocessed one-year after reactor discharge.
- Personnel access to operating areas through hardened tunnels with close control of entry/exit.

- High-level wastes solidified in glass suitable for long-term isolation in a Federal repository.
- Flexibility for changes, additions or upgrade of process equipment, flowsheets, instruments, etc., thus the basic canyon structure can be operated indefinitely.
- No accumulation of separated plutonium except for secure surge storage between reprocessing and fuel fabrication. Plutonium leaving hardened containment (i.e, reinforced concrete structure designed to resist an attack using weapons and explosives) would be in MOX fuel assemblies. A "co-processing" design option would eliminate any separate streams or accumulation of separated plutonium.
- Tritium and krypton removal.
- Sand filtration, because of demonstrated high reliability, long life, high efficiency, high air permeability, inherent freedom from channeling, superior protection during fires, better performance in the presence of moisture, high chemical resistance, self-sealing after disturbances such as earth tremor, tornado or explosion, and ease of maintenance or repair.
- Opportunities for lower cost through research and development, and as a result of the much longer cooling time for spent fuel to be reprocessed.

This facility design concept was not considered in White House reviews of reprocessing during the Ford and Carter Administrations, nor as an option for support by President Reagan, who had been elected on a platform to support reprocessing of commercial spent fuel. The ERDA and the DOE had reassigned responsibilities for commercial fuel cycle to its Division of Reactor Development (later Office of Nuclear Energy) which supported pilot plant concepts of its national laboratories and rejected concepts based on successful experience and lessons learned from that experience.

THE NEW VISION INCLUDES RECOGNITION THAT ACTIVITIES INVOLVING LARGE SCALE USES OF COMPLEX NUCLEAR TECHNOLOGY MUST BE CARRIED OUT BY EXPERIENCED, COMPETENT CORPORATIONS.

General Leslie R. Groves, Director of the US Manhattan Project during World War II, recognized that the difficulties of safe and sustained reprocessing needed for production of plutonium for a nuclear deterrent would be a challenge even to the most experienced chemical corporation. He asked the DuPont Company to design and build the reprocessing pilot plant at the Clinton Laboratories at Oak Ridge, Tennessee, and later to design, build and operate the industrial-scale reprocessing at the Hanford Engineering works in Washington State.[11]

Many of the Manhattan Project scientists were disappointed with decisions to use industrial corporations for reprocessing and other operations for the nuclear project. They believed that their accomplishments had earned them the right to carry the project through to completion, and that time would be wasted in teaching a second group the knowledge that they had already created and mastered. Most of the scientists were young and had no industrial experience or understanding of the difficulties of safe, sustained operations with complex technology.[12]

That vision of Manhattan Project scientists was carried forward in AEC/ERDA/DOE national laboratories, and is documented in a 1994 report by ORNL for the DOE with a statement by a former associate director "It is safe to say that (ORNL) Chem Tech has played the leading role in solving the nation's reprocessing problems. When Alvin Weinberg was ORNL's Director, he used to say that one purpose of the laboratory was to undertake big projects of national purpose that others could not handle. Chem Tech's achievements are testimony to that and have earned the division a lasting place in the history of the country's atomic energy programs."[13]

The vision of the DOE is that national laboratories are competent to carry out activities involving complex nuclear technology. The vision of the US Department of Defense is that "one cornerstone to maintaining quality of the DoD Science and Technology program is the strength of partnerships with industry" and "industry's experience with technology can be essential to quality decisions in managing DoD S&T programs."[14] Hopefully The Congress and The President will keep this in mind when considering a new organization for carrying out important nuclear work of the DOE.

THE NEW VISION INCLUDES A SAFE, EFFICIENT AND RELIABLE PRODUCTION COMPLEX FOR NUCLEAR MATERIALS FOR IMPORTANT NATIONAL PROGRAMS SUCH AS DEFENSE, SPACE EXPLORATION, MEDICINE, INDUSTRIAL NEEDS AND RESEARCH, INCLUDING PRODUCTION AND STUDY OF TRANS-SEABORGIUM ELEMENTS.

The ten years that Glenn Seaborg was Chairman of the USAEC were exciting for managers and workers in nuclear programs, and particularly for nuclear materials production programs at the Savannah River Plant. When he determined, in 1964, that no more weapons plutonium was needed, major efforts were devoted to production of tritium needed to maintain and improve the strategic nuclear deterrent; plutonium-238 for space exploration; higher isotopes of plutonium and uranium-233 for advanced reactor development; high specific activity cobalt for food irradiation and other purposes; and other isotopes for medical and industrial applications and research. Most exciting was adaptation of "C" reactor at SRP for high flux operation for production of kilograms of americium-243 and curium-244; grams of californium-252; milligrams of Einsteinium, Fermium, Mendelevium and Nobelium; micrograms of elements 103 to 108, including element 106 which was named Seaborgium in his honor; and many atoms of higher elements, including elements above 113 that he expected would be more stable.

The "Multi-Purpose Processing Facility" was installed in "F" Canyon (reprocessing plant) at SRP for separation of Californium and trans-californium elements using newly developed, high-pressure, chromatographic cation exchange processes.

The neutron flux achieved at "C" reactor was much higher than that in the High Flux Isotope Reactor (HFIR) at ORNL, and production in heavy water moderated reactors is several orders of magnitude more efficient than accelerators that were used a year ago to produce the first atom of element 114. Dr Seaborg recognized the need for a large, efficient complex for production of higher elements, because of the exponentially decreasing yields with increasing atomic numbers.

All of that capability has been lost by the Department of Energy, including the industrial contractor that was responsible for its success.

The tritium that is needed to maintain the nuclear deterrent will be produced in commercial nuclear power plants which is inefficient and a compromise of important and long standing non-proliferation practice; plutonium-238 needed for space exploration is being purchased from Russia.

Hopefully The Congress and The President will keep the need for this vision in mind when considering a new organization for carrying out all nuclear work of the Department of Energy.

THE NEW VISION INCLUDES VIABLE PROGRAMS FOR MANAGEMENT OF THE BY-PRODUCTS OF THE NUCLEAR FISSION PROCESS, INCLUDING RECYCLE OF ALL FISSIONABLE MATERIALS AND LONG-TERM ISOLATION OF UNWANTED FISSION PRODUCTS FROM THE BIOSPHERE.

In May, 1991, Henry Thomas, a former Assistant Secretary of Energy and executive of the American Gas Institute, described his vision for an energy future in the US to participants at a Federal executive seminar on foreign policy. He described nuclear power as safe - but then said that natural gas would become the energy resource of choice because of the inability to manage nuclear wastes.

Mr. Thomas's vision is sound, except that the problem is not inability, but unwillingness to manage and dispose of wastes from commercial nuclear power plants and recycle valuable materials.

The DOE draft Environmental Impact Statement for the Disposal of Spent Nuclear Fuel and High-level Radioactive Waste at Yucca Mountain in Nevada was issued in July 1999. The action described is appropriate for disposal of radioactive waste, i.e., the fission products remaining after recovery of fissionable materials in reprocessing. However, it is not appropriate for disposal of the plutonium and other weapons or weapons source material in spent fuel, that become accessible for easy recovery with decay of thirty-year half-life cesium-137. The action also denies use of an energy resource that could be used to supply all of the electricity needs for the United States for 10,000 years, without release of atmospheric pollutants or greenhouse gases that threaten catastrophic climatic changes. Thus there is no program nor plan for a program for appropriate disposal of commercial nuclear wastes.

Some utilities are providing long-term on-site dry cask storage; others are negotiating with Native American nations for storage on reservations, while others propose shipment of spent fuel to Yucca Mountain for interim, retrievable storage. But these options do not address waste disposal and will not convince an informed public that the waste problem has been solved.

Hopefully The Congress, The President, and leaders of nuclear utilities will keep the need for this vision in mind during deliberations for new organizations that will carry out important nuclear work of the DOE, and will decide to create a new corporation under the direction of nuclear utilities that will begin programs for management of commercial spent fuel. The DOE should make available to this corporation all funds collected under the Nuclear Waste Policy Act (NWPA), plus interest, and at least two sites, such as the SRP in South Carolina and the Hanford Works in Washington State.

NWPA funds should be supplemented by amounts equivalent to those now devoted to environmental restoration at the sites. Nuclear utilities should request assistance from experienced corporations in setting up the new corporation for spent fuel management and carrying out this important work, and ensure that best technology is used.

THE NEW VISION INCLUDES AN AMERICAN PUBLIC THAT IS PROVIDED FULL AND ACCURATE INFORMATION ON THE SAFETY, ENVIRONMENTAL ADVANTAGES, MEDICAL USES, AND OTHER BENEFITS OF NUCLEAR TECHNOLOGY; THE LACK OF DANGERS FROM WELL-MANAGED NUCLEAR TECHNOLOGY; AND THE CONSEQUENCES OF POOR MANAGEMENT AND EFFORTS TO ASSURE THAT IT IS WELL MANAGED.

This is a vision of Glenn Seaborg. Soon after he became Chairman of the AEC, this vision was initiated in portions of the AEC. The program was continued under Chairmen James Schlesinger and strengthened by Dixy Lee Ray. Richard Roberts, the first Assistant Administrator for Nuclear Energy in the Energy Research and Development Administration (ERDA) attempted to continue the program, but was ordered by the ERDA Office of General Counsel not to do so, and this order has been sustained under the DOE.

The Institute for Energy and Environmental Research (IEER), a well-organized and well-funded anti-nuclear organization in Takoma Park, Maryland, publishes *Science for Democratic Action (SDA)* four times a year to "provide the public and policy-makers with thoughtful, clear, and sound scientific and technical studies on a wide range of issues. IEER's aim is to bring scientific excellence to public policy issues to promote the democratization of science and a healthier environment."[15]

SDA is well written, but explains with false or misleading, often inflammatory information why everything being done with nuclear technology is wrong and should be phased out, all waste placed at the Waste Isolation Pilot Plant should be removed, excess Russian and US weapons plutonium should not be destroyed by its use as MOX, and glassification of high level radioactive waste should be replaced by calcination whose product is not only soluble but hygroscopic.

SDA is distributed to local public interest groups; news media; local, state and national political representatives; and others. Major newspapers have adopted ideas from *SDA* for editorial policies and rely on Arjun Makhijani, IEER President, as an authority on nuclear technology. The lead article in the Massachusetts Institute of Technology's August/September 1988 issue of *Technology Review* by Mr. Makhijani and Robert Alvarez, later DOE Deputy Assistant Secretary for Policy, featured false allegations of dangers of DOE nuclear waste, including probability of a Chornobyl-scale explosion. *The Washington Post* article, "Nuclear Waste: The $100-Billion Mess" was adapted from the *Technology Review* article for the entire front page of the "Outlook" section of its September 4, 1988 issue. The DOE had a comprehensive report from The DuPont Company[16] refuting the allegations of danger and thus knew that they were false, but made no effort to correct the misinformation. *Technology Review* published a letter to the editor[17] correcting the misinformation, and later the editor-in-chief admitted that the article had been a serious mistake. *The Washington Post* did not publish any correcting information. This misinformation led to increased public fears of non-existing dangers and appropriation of approximately $6 billion per year for "Nuclear Waste Cleanup" beginning in 1990. Tens of billions of dollars have been wasted.

In June 1995, 43 individuals and representatives of organizations who have worked to develop the peaceful uses of nuclear science and technology signed a declaration of interdependence in support of the Eagle Alliance. In the charter they acknowledged existence of influential interests who strive to deny others access to nuclear technology, and who have occupied the domain of public opinion by the restraint and silence of knowledgeable individuals and organizations.

However, the corporation formed to implement this declaration of interdependence has failed to provide information to the public, news media, local, state and national political representatives, or others. There will be no level playing field for use of nuclear technology unless there is a well-coordinated national effort to provide accurate information to the public and to correct misinformation.

SUMMARY AND CONCLUSION

The great benefits of nuclear technology will not be fully realized until the nuclear community adopts a new vision, which includes:

- Full cooperation among nations for peaceful uses of nuclear technology
- Assurances that best technology will be available for peaceful uses of nuclear technology by all nations, and that information on successful and unsuccessful experiences will be provided
- Recognition that activities involving large scale uses of complex nuclear technology must be carried out by experienced, competent corporations
- A safe, efficient and reliable production complex for nuclear materials for national programs such as defense, space exploration, medicine, industrial applications and research - including the production and study of trans-Seaborgium elements
- Viable programs for management of the by-products of the nuclear fission process, including recycle of all fissionable materials and long-term isolation of unwanted fission products from the biosphere
- An American public that is provided full and accurate information on the safety, environmental advantages, medical uses, national and global security, and other benefits of well-managed nuclear technology.

REFERENCES

1 Executive Office of the President, Office of Science and Technology Policy, *Climate Change: State of Knowledge* (October 1997) This document summarizes conclusions of the 1995 report of the Intergovernmental Panel on Climate Change, the most comprehensive and thoroughly reviewed assessment of climate change science ever produced, representing the work of more than 2,000 of the world's leading climate scientists.

2 Mobil Corporation, *Climate: Technology and Carbon Dioxide Emissions: A Global Review and Assessment*, (1999)

3 Nakicenovic, et al *Global Energy Perspectives*, IEA (1998)

4 Institute for Energy and Environmental Research, *Science for Democratic Action*, (July 1999). Statement in masthead of publication, page 2.

5 Bastin, Clinton, letter to Naresh Chandra, India's Ambassador to the United States (June 12, 1998) Copies of that letter and replies from Ambassador Chandra and the US National Security Council can be obtained from the author.

6 USAEC Division of Civilian Application Summary Report: *AEC Reference Fuel-Processing Plant*, WASH-743 (October 1957)

7 ibid, page 1

8 AEC official accountability records of enriched uranium shipped from ICPP, reviewed by the author in 1973.

9 Low, Lawrence D., Director, Division of Compliance, US Atomic Energy Commission, letter to R. N. Miller, President, Nuclear Fuel services, Incorporated (March 16, 1972)

10 Reed, C. E., Senior Vice President - Corporate Studies and Programs, et al, General Electric Company *Midwest Fuel Recovery Plant Technical Study Report* (July 5, 1974)

11 Hewlett, Richard G, and Anderson, Oscar E., Jr., *The New World - 1939/1946 (Volume I of a History of the United States Atomic Energy Commission)*, page 91.

12 Hewlett, Richard G, and Anderson, Oscar E., Jr., *The New World - 1939/1946 (Volume I of a History of the United States Atomic Energy Commission)* and Rhodes, Richard, *The Making of the Atomic Bomb*.

13 Jolley, Robert L, Genung, Richard K., McNeese, L.E. (Gene), and Mrochek, John E., *The ORNL Chemical Technology Division: 1950-1994*, page 1-23.

14 Etter, DeLores M. (Deputy Under Secretary of Defense, Science and Technology), letter to Clinton Bastin, (June 22, 1999)

15 Same as reference 4.

16 DuPont Company Report to the Department of Energy "Response to Environmental Policy Institute Report on Savannah River Plant High-Level Waste Management," DPSP-86-1164, (December 31, 1986)

17 Bastin, Clinton, Letter to Editor, *Technology Review* (April 1989) This letter included information from the DuPont Company report (Reference 13). A "response from the authors" was also published. Note that the authors misquoted from the letter in order to refute the claims of misinformation.

NUCLEAR POWER IN THE CONTEXT OF CRITICAL GLOBAL PROBLEMS*

David Bodansky

Department of Physics
University of Washington
Seattle, WA 98195

INTRODUCTION

There is now a marked pause in the construction and deployment of new nuclear power plants. Although some construction of reactors continues in Asia and Eastern Europe, a de facto moratorium exists in the United States and in most of Europe, while in Sweden and Germany the governments plan to shut down operating plants before the end of their normal lifetimes. The inhibitions on nuclear power development stem in large measure from environmental concerns, particularly concerns relating to reactor accidents and nuclear wastes.

Unfortunately, stopping the development of nuclear power may do more environmental harm than good. We consider below a number of global problems – with important connections to nuclear power – that involve environmental risks of greater size and global scope than those posed by nuclear reactors or nuclear wastes.

To emphasize the difference in scale, the different classes of problems are here classified as *confined* problems and *open-ended* problems. Confined problems are those where the probability and magnitude of the risks can be quantitatively studied and are found to be limited in scope. Reactor safety and nuclear waste disposal are in this category.

In contrast, for the open-ended problems it is difficult to evaluate the magnitude of the potential consequences, but in plausible scenarios they may involve great harm on a global scale. These problems include:

- The danger that nuclear explosives may be used by warring nations or terrorists.
- The gradual exhaustion of resources of oil, natural gas, and coal.

---

*This paper is based in part on a paper presented at the Centennial Meeting of the American Physical Society (Atlanta, March 1999); a condensed version of that paper appears in Bodansky, 2000.

- The possibility that increased atmospheric carbon dioxide levels will cause major adverse climate changes.
- The difficulty of meeting the increasing energy demands of a world population that is growing in size and material aspirations.

These matters all involve high stakes, with the possibility of impacting many millions of people.

CONFINED PROBLEMS

Nuclear reactor accidents

The belief that risks from reactor accident are small is based on the past safety record of nuclear reactors, results of on-going probabilistic risk analyses, indicators of improvement in reactor performance, and the prospect of still greater safety in a next generation of nuclear reactors.

The past safety record of nuclear reactors, other than the Soviet Chernobyl-type RBMK reactors, is excellent. Excluding RBMK reactors, there had been about 9000 reactor-years of operation in the world by the end of 1999, including about 2450 in the United States.[1] In this time there was only one accident involving damage to the reactor core, the 1979 Three Mile Island accident, and even at TMI there was very little release of radionuclides to the outside environment.

There was a very large release of radionuclides in the 1986 Chernobyl accident. The explosion and radiation exposures led, in a period extending from the day of the accident to a few months later, to 31 deaths among workers at the plant. Fourteen additional workers died over the next ten years (IAEA, 1996). Beyond this, the only clearly established additional health effect has been a large increase in the rate of thyroid cancer among children living in the vicinity, including 3 fatalities (as of 1996). It is estimated that the cumulative radiation exposure of the population of the Northern Hemisphere may eventually cause up to 30,000 additional cancer fatalities during the 70-year period following the accident. However, the average exposures were small and it is possible that the calculation method – based on the so-called *linearity hypothesis* – overestimates the actual consequences. There is little prospect of observing this increase, because these cancers will be overshadowed by the 800 million natural cancers expected during the same period.

Although the Chernobyl accident was very serious, the defects in design and operating procedures that led to it were so egregious that the accident has little relevance to current reactors outside the former Soviet Union. However, it serves as a reminder of the need for rigorous care in the design, construction, and operation of nuclear power plants.

The TMI and Chernobyl accidents both had large negative impacts on public attitudes towards nuclear power. Further reactor accidents of any sort, even one that causes as little harm to health as TMI, will make it all the harder to revive nuclear power in Europe or the United States. The past reactor accident experience, along with the grossly negligent behavior that led to a fatal accident at the Tokaimura fuel conversion facility in Japan in August 1999, brings home the importance of making a next generation of reactors as immune to human carelessness or ignorance as possible. This is especially urgent if nuclear power is to be used in countries that have not yet achieved the level of technological

[1] Extrapolated from data in IAEA, 1998.

sophistication and regulatory rigor that has now been reached in, for example, the United States.

Partly with the high stakes in mind, changes have been made in U.S. reactor equipment and operation since the TMI accident to reduce the chance of another accident. The results of these changes are reflected in the predictions of probabilistic risk assessments and by a variety of direct performance indicators. For example, in one measure for U.S. reactors, since the pre-TMI days there has been a reduction of more than a factor of 100 in the number of precursors to potential core damage accidents, as reported by the Nuclear Regulatory Commission (Murley, 1990; Belles *et al.,* 1998).

A next generation of reactors can be even safer, either through a series of relatively small evolutionary steps that build directly upon past experience or through more radical changes that place greater reliance on so-called "passive" safety features. A reasonable target for a next-generation reactor might be to keep the chance less than one in a million per year for an accident involving core damage and less than one in ten million per year for an accident that leads to a large external release of radioactive material. In a world with roughly ten times as many reactors as today, say 4000 reactors, this would correspond to a 4% chance per century of a reactor accident leading to a significant release of radionuclides, i.e. something like one chance in twenty-five of "another Chernobyl" per century.

If this safety level is achieved, nuclear reactors would already be relatively benign neighbors, but in fact reactor manufacturers expect to do even better. For example, the General Electric Company estimates that for its evolutionary Advanced Boiling Water Reactor – two early versions of which are now in operation in Japan – the core damage frequency is less than 2 in 10 million per year (GE 1999). The estimated large release probability is a factor of 500 smaller. Of course, manufacturer's expectations must be examined with an independent critical eye. But it is reasonable to expect that the nuclear industry will have learned from past experience, both good and bad, how to make a new generation of reactors substantially safer than the impressively safe previous generation,

Nuclear waste disposal

A second major public concern is over nuclear wastes. Most experts believe that it is possible to dispose of these in a manner that poses little threat to the environment and human health, given the small volume of the spent fuel, the decay with time of the radionuclides, and the potential effectiveness of engineered and natural barriers. The success that is likely to be achieved is examined through Total System Performance Assessments (TSPA) (see, e.g., OCWRM, 1998).

Without confronting the complexity of studying and evaluating the TSPAs, one can gain some perspective on the scale of the hazards by considering the protective standards that have been proposed for nuclear waste repositories, in particular for the proposed U.S. site at Yucca Mountain (Bodansky, 1996). There have been three major proposals in recent years:

- *EPA proposed standard in 40CFR191 (EPA, 1985).* Releases of radionuclides from the repository were to be limited to amounts such that the projected number of premature cancer fatalities over a 10,000 year period would not exceed 1000, i.e. an average of one per decade.[2] This target appeared to be attainable until it was

[2]The specified limit is for a repository containing 100,000 tonnes of fuel. For Yucca Mountain, with an expected 70,000 tonnes, the limit would be proportionately lower.

recognized that the relatively rapid movement of gases through the site would allow the escape of carbon-14, in the form of carbon dioxide, in amounts that could raise the global concentration of carbon-14 in the atmosphere by about 0.1% above its natural level. With a strict application of the linearity hypothesis to the resulting very small incremental doses, the EPA calculations indicated that the release of the entire carbon-14 inventory could cause 4000 worldwide cancer deaths over 10,000 years.[3] Congress subsequently suspended the EPA's authority over Yucca Mountain, pending recommendations of a Committee to be appointed by the National Academy of Sciences (NAS).

- *National Research Council committee recommendations (NAS/NRC, 1995).* The NAS report mandated by Congress recommended that the standard be based on the average exposure to a small "critical group" (representative of those receiving the highest exposures) and that the average exposure for members of this group not result in an individual risk of a fatal cancer exceeding 10^{-5} or 10^{-6} per year, for a period of up to one million years. If present conventional dose-response relations are used, this corresponds to annual doses of 0.2 mSv or 0.02 mSv, respectively.[4]
- *EPA proposed standard in 40CFR197 (EPA, 1999).* The standard here is based on the dose to the "reasonably maximally exposed individual." The proposed limit is set at 0.15 mSv per year for the next 10,000 years. The U.S. Nuclear Regulatory Commission (NRC) has made numerous specific criticisms of the EPA proposal, including suggestions that the dose limit be raised to 0.25 mSv per year (Travers, 1999). Such criticisms, and inputs from other sources, may significantly delay the promulgation of the final standard.

The array of proposed standards differ in details, but the approaches are the same in two important ways: (a) no account is taken of possible technological or medical advances during the next millennia; (b) the level of harm to be avoided in the distant future is miniscule compared to the level of harm that society accepts with a shrug today – for example, the dose of about 2 mSv per year that the *average* person in the U.S. now receives from indoor radon, with projected lung cancer fatalities in excess of 15,000 per year.

It should be noted that there is intense controversy as to the health effects of radiation doses below about 100 mSv per year. This estimate of 15,000 annual cancer deaths from indoor radon, as well as estimates of tens of thousands of eventual cancer deaths from Chernobyl exposures, is obtained by applying the linearity hypothesis. This hypothesis has been adopted by most regulatory agencies but is strongly contested by some scientists who believe it overestimates the effects of radiation at low dose levels. Of course, if calculations based on this hypothesis overestimate the deaths from indoor radon, they also overestimate the effects of potential radiation from a waste repository.

Overall, in focussing on such quantitatively minute and temporally remote harm, the NAS panel, the EPA, and the NRC are providing very suggestive, albeit indirect, evidence that the dangers from Yucca Mountain are small – certainly small compared to the open-ended dangers considered below. Indeed, it is hard to avoid the impression that the concern

---

[3]The same approach implies 50,000,000 calculated cancer deaths from *natural* carbon-14 in the atmosphere for a population of 10 billion over a 10,000 year period.

[4]1 millisievert (mSv) = 10^{-3} sievert (Sv) = 100 millirem (mrem) = 0.1 rem. The average annual radiation dose in the United States from natural sources is about 3 mSv per year, including about 0.01 mSv from natural carbon-14 in the atmosphere.

about small exposures 10,000 years hence is something of a ritualized exercise, designed more to forestall criticism than to protect future populations.

This impression is reinforced by the absence of an attempt to consider the context in which the hazards might plausibly arise. The most probable future world is one that has advanced far beyond us in medicine and technology, making irrelevant a concern over small exposures from radiation. A much grimmer alternative would be a world with an intervening "dark age," leaving societies whose scientific and technological capabilities are more primitive than our own. If we think such a disaster is plausible, it behooves us to focus on measures that would help avoid it. For the relatively near term, this would include steps to reduce the likelihood of global warfare or global economic collapse.

In summary, if one adopts the default assumption of no change in society, then small exposures from Yucca Mountain are of minor interest. If one assumes either of the changed futures sketched above, they become a complete irrelevancy. Our real responsibility is not to guard against relative trivia, but rather to make sure that there is virtually no possibility of large doses to large numbers of people. It appears highly unlikely that this could occur, but it would be helpful if government standards, and associated studies, were focussed on establishing and verifying this criterion. Defining the meaning of "large" would be a difficult and contentious matter, but it would place the debate in an area of appropriate concern.

OPEN-ENDED PROBLEMS

Nuclear weapons proliferation

The first of the open-ended problems to be considered is nuclear weapons proliferation, in the context of its relation to commercial nuclear power. There is a connection, because a country with an active nuclear power program has a head start, in terms of equipment and technically trained people, should it decide to embark upon a weapons program. This has been a live issue in the case of Iran.[5]

Historically, however, commercial nuclear power has played little or no role in nuclear weapons proliferation. The long-recognized nuclear weapons states – the United States, the Soviet Union, the United Kingdom, France, and China – each had nuclear weapons before they had electricity from nuclear power. India's weapons program was initially based on plutonium from research reactors and Pakistan's on enriched uranium, although this does not rule out the possibility of later linkages between their weapons and civilian programs. The three other countries whose suspected nuclear weapons programs have attracted the most recent attention – Israel, Iraq, and North Korea – have no civilian nuclear power whatsoever. Further, many countries started their weapons programs with uranium-235 as the fissile material, not plutonium-239 as would be the case in the usual proliferation scenarios.[6]

For the United States to abandon nuclear power would not help to thwart potential proliferation unless at the same time we would relinquish our nuclear weapons and could stimulate a broad international taboo against all things nuclear. Clearly, we have no will to do this. Further, whatever policy the U.S. were to adopt, over 30 countries now use

---

[5]The United States has opposed the aid that Russia is giving to Iran in building nuclear reactors for electricity. This is in contrast to the aid the U.S. has offered to North Korea for a similar reactor program.

[6]Highly enriched uranium-235 was used for the U.S. bomb dropped on Hiroshima, for the first bomb tested by China, for the first weapons efforts of Pakistan, and for the since-abandoned weapons programs of Argentina, Brazil, and South Africa.

nuclear power for generating electricity and many more have research or test reactors. A comprehensive nuclear taboo is highly unlikely, given the heavy dependence of France, Japan, and others on nuclear power, the importance of medical uses of radionuclides, and the wide diffusion of nuclear knowledge among countries that differ greatly in their sense of political morality, their economic options, and their perceived military pressures.

A more promising approach lies in stringent control and monitoring of commercial nuclear power programs, such as attempted by the International Atomic Energy Agency. The U.S. voice in the design of future reactors and fuel cycles and in the shaping of the regulatory regimes that might govern them is likely to be stronger if the United States remains a leading player in the civilian nuclear power enterprise

Further, the threat of future wars may be diminished if the world is less critically dependent on oil. Competition over oil resources was an important factor in Japan's entry into World War II and in our military response to Iraq's invasion of Kuwait. Nuclear energy can contribute to reducing the urgency of such competition, albeit without eliminating it.

Finally, there is the risk that terrorist groups could steal potential bomb materials. Such concerns influenced the U.S. decision in the late 1970s to abandon the reprocessing of spent fuel.[7] This lead has not been followed elsewhere, with France, India, Japan, Russia, and the United Kingdom continuing reprocessing. This issue may become crucial if nuclear power is to continue operation far into the future and breeder reactors are found to be essential. There is no immediate pressure to move to breeder reactors, because present uranium supplies could accommodate a large nuclear expansion for many decades.[8] Nonetheless, for the long-term it is important to develop fuels cycles that exploit more of the energy potential of uranium or thorium without increasing the opportunities for the diversion of fuel to bombs.

In summary, none of the links between nuclear power and nuclear weapons appears to be very strong, and even the net direction of the possible coupling is in doubt. The use of nuclear weapons involves such major consequences that the surrounding issues must continue to be central to the consideration of nuclear power. It is not clear whether consideration of these dangers will provide a better argument for nuclear power or against it. But assuming nuclear power undergoes a major expansion, weapons proliferation concerns should enter strongly into the design and operation of the future nuclear fuel cycles.

The depletion of fossil fuels

The United States and the world are overwhelmingly dependent on fossil fuels as the main source of energy. As of 1997, fossil fuels provided 86% of U.S. primary energy and 86% of world primary energy. The era of fossil fuels began some 200 years ago with heavy use of coal in Great Britain and it is almost sure to come to an end over the next 100 or 200 years as the remaining resources of oil, natural gas, and coal are consumed and other energy sources become dominant.

---

[7]The fuel removed from reactors contains large amounts of plutonium. The large plutonium-240 admixture in "reactor-grade" plutonium makes it difficult to build an effective bomb, but not impossible to do so. The very high level of radioactivity of the spent fuel provides a protective barrier against diversion of the fuel by any group that lacks extensive facilities for handling and transporting the fuel, greatly complicating potential clandestine diversion attempts.

[8]If one assumes resources of 20 million tonnes of uranium and a requirement of 200 tonnes/GWyr, affordable resources would suffice for about 100,000 gigawatt-years of light water reactor operation (Bodansky, 1996, Sec. 7.5).

There is some dispute among analysts as to whether world production of conventional oil will peak before the year 2020 or whether the peak will be delayed by another decade or two (Kerr, 1998), but in either case the current era of relatively cheap oil will end within several decades. A similar scenario is likely to follow for natural gas, although at a slower pace, and at a still slower pace, for coal. If our responsibilities to future generations include the relatively small problems that nuclear waste repositories may create in 10,000 years, they also include preparing for fossil fuel scarcity that will occur very much sooner.

Nuclear fission is not the only means towards addressing the matter. Clearly, there is an important further role for conservation. Renewable sources can also make a contribution, but some caution is suggested by the inverse correlation that now exists between the extent to which a renewable source is used and the degree that an open-ended expansion is possible. In particular, hydroelectric power and geothermal power have clear limits on their expansion, biomass and wind have somewhat ambiguous expansion possibilities, and photovoltaic and solar thermal power – with their apparently open-ended potential – still make only very small contributions and the practicality of their large-scale expansion remains to be demonstrated.[9] Fusion offers the prospect of a major new source, but at best it is many decades away and at worst it may never prove to be practical. Overall, it would be a tremendous gamble to assume that fossil fuels can be replaced without the use of fission energy.

Global climate change

Quite apart from limits on resources, the prospect of global climate change caused by increases in the atmospheric concentrations of greenhouse gases suggests that we should speed the replacement of fossil fuels. As discussed, for example, in the reports of the Intergovernmental Panel on Climate Change (IPCC), there is a significant possibility of large, and on balance harmful, effects from the increased concentrations of carbon dioxide in the atmosphere (IPCC, 1995). In the central, most probable, IPCC projections for the year 2100, the average temperature increase is about 2 °C and the sea level rise is 50 centimeters – but considerably higher (and lower) values are not excluded. These increases are only the beginning, and further changes would continue beyond the year 2100 unless carbon dioxide emissions are curtailed before then. If one were to follow the example of nuclear energy, it would be appropriate to consider events of low probability and high consequences. The possible collapse of the West Antarctic ice sheet, leading to an estimated sea level rise of 5 meters, would fall into this category.[10]

We will not explore the predicted effects further, but note that most governments profess to take them seriously, and, more significantly, most atmospheric scientists take them seriously. The United States agreed under the Kyoto Protocol to bring carbon dioxide emissions in the year 2010 to a level that is 7% lower than the 1990 level (DOE/EIA 1998). Given the intervening increases, this target is 16% lower than the actual 1998 level. It will be difficult to achieve. This short-term target may in itself not be urgent, except perhaps

---

[9]Wind, solar thermal, and photovoltaic power together accounted in 1998 for only about 1% of U.S. electricity generation from renewable sources and only about 0.1% of all electricity (DOE/EIA 1999a).

[10]John Houghton, who served as co-chairman of the Scientific Assessment Work Group of the IPCC assesses the hazard in the following terms, offering reassurance only for the short-term: "Although scientists are not yet very confident in their ability to model the dynamic behavior of large ice-sheets, there is no reason to suppose there is a danger in the short term (for instance, during the next century) of collapse of any of the major ice-sheets. Much greater understanding of the behavior of large ice-sheets must be obtained before the amount of warming which might induce such collapse can be estimated" (Houghton, 1997, p. 110).

for symbolic purposes, but the broad goal of restraining carbon dioxide emissions through the next century remains important.

The use of coal for electricity generation is responsible for about 32% of anthropogenic carbon dioxide emissions in the U.S.[11] As shown by France, it is possible to displace virtually all the coal used in electricity generation. Thus, France in 1997 obtained about 78% of its electricity from nuclear power and only about 5% from coal. Further reductions in carbon dioxide emissions could be made by the electrification of other sectors of the energy economy, including buildings, and eventually perhaps much of transportation.

Again, as in finding substitutes for fossil fuels, mitigation approaches include conservation, renewable energy, nuclear fusion energy, and nuclear fission energy – with the same caveats as above. A switch from coal to natural gas is also an effective means for reducing carbon dioxide emissions, although this is almost literally a half-measure. Sequestration of carbon dioxide, either in vegetation through photosynthesis or by capture of escaping gases from power plants, is another possibility, but with uncertain scope and practicality. Overall, restraining the growth of carbon dioxide in the atmosphere will be difficult, and prudence suggests that all promising options be explored and exploited.

Global population growth and energy limits

The problem of finding energy sources to replace fossil fuels will be made more difficult by the increase in energy demand caused by the growth of world population and by the higher economic aspirations of most of this population. The world population was 2.5 billion in 1950 and has risen to about 6 billion today. It seems headed towards 10 billion, and perhaps beyond, in the next century. This growth will inevitably come up against the obstacle of limited energy supplies.

The broad problem of resource limitations in the face of a rising population is sometimes couched in terms of the "carrying capacity" of the Earth, or alternatively as the question that provides the title of the very comprehensive 1995 book by Joel Cohen, *How Many People Can the Earth Support?* (Cohen, 1995). Defining what one means by carrying capacity involves both practical considerations and one's sense of values. Thus as Cohen writes:

> If an absolute numerical upper limit to human numbers on the Earth exists, it lies beyond the bounds that humans would willingly tolerate (*ibid*, p. 359).

This is put perhaps even more succinctly by Garrett Hardin, in a review of Cohen' s book:

> What one really wants to know is this: after we define the minimally rich sort of life we human beings would consent to live, what is the maximum number of people possible (Hardin, 1996).

Cohen reviews a large number of attempts to estimate the world's ultimate carrying capacity, dating back to the Dutch naturalist Antoni van Leeuwenhoek in 1679. As summarized by Cohen, Leeuwenhoek estimated the population density of Holland to be 120 per km^2, assumed that land encompassed one-third of the Earth's total area, and extrapolated to a world population of about 13 billion.[12]

---

[11] For 1998, U.S. CO_2 emissions were: 32% from coal in electricity generation; 5% from oil and gas in electricity generation; 33% from oil (and a small amount of gas) in transportation; and 30% from fossil fuel use in industry, commercial activities, and residences (DOE/EIA 1999b).

[12] This sort of analysis embodies the so-called "Netherlands fallacy." The fallacy lies in ignoring the dependence of densely populated areas on imports from less densely populated areas.

Remarkably, more recent estimates of the Earth's carrying capacity center around a value of about 10 billion, not meaningfully different from the three-century old estimate of Leeuwenhoek. However, the range of estimates is very great – from under 2 billion to well over 20 billion. If one performed today a calculation analogous to that of Leeuwenhoek, based on present national population densities, the world population extrapolates to about 3.6 billion with the United States as the reference base and 31 billion with the United Kingdom as the reference.[13]

The constraints that limit world population can be put in several categories:

a. *Material.* Population is limited by the accessible supply of necessities and of desired amenities, from food to parking places. Key material factors include land area, energy, and water.

b. *Ecological.* Growth in human population places strains on the overall environment, including destruction of wilderness and extinction of other species.

c. *Aesthetic or philosophical.* These constraints are suggested, for example, in a quotation from John Stuart Mill, cited by Cohen:[14]

> A population may be too crowded, though all be amply supplied with food and raiment. It is not good for man to be kept perforce at all times in the presence of his species....Solitude, in the sense of being often alone, is essential to any depth of meditation or of character; and solitude in the presence of natural beauty and grandeur, is the cradle of thoughts and aspirations which are not only good for the individual, but which society could ill do without.

This was written in 1848 when the world population was about one billion.

Arguments based on (b) and (c) are difficult to quantify and put in "objective" terms, although in fact they may the most emotionally compelling of all. It is perhaps for this reason that most of the stated rationales for a given carrying capacity are based on material arguments. The most critical material constraint is that of food supply, which in turn depends upon arable land area, energy, and water. In particular, energy is required for irrigation, the production of fertilizers, the operation of farm machinery, and the transportation of farm products.

Carrying capacity estimates made directly in terms of energy, in recent papers by David Pimentel and collaborators (1994) and by Paul Ehrlich and collaborators (Daily, 1994) can serve as illustrations of the possible implications of restricted energy supply. Each group concludes that an optimal global population for a sustainable future is under 2 billion – a much smaller limit than given in most other estimates. The argument is made most explicitly in the Pimentel paper. The authors envisage a world in which solar energy is the only sustainable energy source. They take 35 quads of primary solar energy to be the maximum that could be captured each year in the United States. Assuming that the present average per capita U.S. energy consumption is halved through conservation and energy efficiency, the 35 quads would suffice for a population of 200 million. For the world as a whole, the total available energy is estimated to be about 200 quads. If the world per capita energy consumption were to converge to the hypothetical future U.S. average (one-half the present U.S. average) this would support a population of somewhat over 1 billion, which Pimentel *et al.* interpret as meaning that "1 to 2 billion people could be supported living in relative prosperity."[15]

---

[13]Using the UK as a reference of course repeats the Netherlands fallacy (see above). These are not extreme reference cases: Bangladesh extrapolates to a population of about 100 billion and Australia to 0.3 billion.

[14]From *Principles of Political Economy* by John Stuart Mill, as quoted by Cohen (1995), p. 397.

[15]A population of 2 billion would correspond to an annual per capita consumption rate of 100 MBTU, compared to the rate of 175 MBTU projected for the U.S. [1 MBTU = 10^6 BTU; 1 quad = 10^{15} BTU.]

One can quarrel with the details of this argument, including the maximum assumed for solar energy, the casual dismissal of nuclear fission and nuclear fusion, and the assumption of a much-improved standard of living for most of the world population. Nonetheless, it illustrates the magnitude of the stakes, and the centrality of energy considerations. It is difficult to believe that the world population could shrink to 2 billion from its present 6 billion – and possible future 10 billion – without great social upheavals.

An increase in energy supplies would obviously relieve the pressures. While major additional contributions might come from renewable energy or fusion, it would be imprudent to count on them. Therefore, to avoid a tremendous gamble on the economic and social future of the world, it is important to lay the foundation for a substantial expansion in the use of fission energy.

It is neither possible nor necessary to know at this time how far this expansion is to proceed. France, with a population of 60 million, obtains about 4 quad of energy per year from nuclear power---a per capita rate of about 67 MBTU per year. This roughly equals the present average world rate of primary energy consumption from all sources and represents a significant fraction of a possible future world average.[16] Were France's example of nuclear use to be widely emulated, constraints on uranium supplies would eventually force the adoption of fuel cycles that use uranium (or thorium) more efficiently – such as breeder reactor cycles. But even with present-day light water reactors, uranium resources suffice for a large increase in nuclear power use.

To illustrate the options that are made available when there is an ample energy supply, we will consider a rather extreme case – the desalination of seawater. It would make little sense to undertake this in the United States in the predictable future, except in limited local situations, because we are not faced with imminent national water shortages, but other countries have little alternative to desalination and even in the United States there are already some desalination projects.

To produce one cubic meter of water in large-scale reverse osmosis plants is expected to cost about $1 and require 6 kilowatt-hours of electricity (Kupitz, 1995; Breidenbach, 1997). Per capita water usage is a little over 2000 cubic meters per year in the United States – about three times the world average and more than twice the average for Europe and Japan.[17] Thus the plausible need for desalinated water in the United States is well below an average of 1000 cubic meters per year per person. At this unrealistically high use rate, each person's share of the national water budget would be about $1000 – showing up in large measure in indirect ways including higher food and electricity costs.[18] To provide this electricity would mean, on a national basis, a 50% increase in total generation. This is a sizable increase, but again not a prohibitive one. It could be accomplished over 40 years with a 1% annual increase in generation per capita (everything else remaining equal).

These numbers do not provide an argument for desalination on anything like the hypothesized scale, much less an argument for nuclear energy per se. But they provide an illustration of the ways in which having ample energy supplies can help to ease the support of a larger world population.

---

[16]Per capita annual energy consumption rates in 1996 averaged about 400 MBTU for the United States, 200 MBTU for France, under 40 MBTU for China, and 80 MBTU for the world. For some developing countries this does not include contributions from wastes and other forms of biomass.

[17]This is the total rate of water withdrawal from all sources and for all purposes, divided by the total population (Gleick, 1993, Section H).

[18]As a point of comparison, we can note that in 1981, when oil prices were at their peak, the average per capita expenditure for motor gasoline and other petroleum products was over $2000, expressed in 1998 dollars (DOE/EIA 1999a). In 1995, it was under $1000, illustrating the size of "acceptable" swings.

Easing the way to a large population is of course not an unmixed blessing. As already suggested there are strong ecological, aesthetic, and philosophical objections to having large further increases in population. But it is inappropriate to use energy limitations as an indirect means for forcing population limits, if these energy limitations are avoidable.

SUMMARY AND CONCLUSIONS

Comparison of the confined and open-ended problems

In evaluating options for obtaining the energy needed to sustain world economic progress in the coming years, it is important to consider the full spectrum of risks to the environment and to human health that each option may create or reduce. This is particularly important in evaluating nuclear power, where often attention has been disproportionately focussed on the presumed dangers.

It has been argued in the preceding paragraphs that the risks from nuclear reactor accidents and nuclear waste disposal are limited in scope and with adequate care can be made small, while, in contrast, the world faces major problems connected with climate change, nuclear weapons, and a mismatch between world population and world energy supply. The contrasting of these classes of problems has so far been in terms of qualitative descriptions, such as "confined" and "open-ended" or "limited" and "major." Difficulties arise in attempting to make these definitions and assignments more quantitative, partly because the boundaries are a matter of subjective opinion and partly because the consequences cannot be known with sufficient certainty. Nonetheless, it is probably appropriate to suggest the sorts of numbers that motivate these qualitative designations.

As discussed above, in a world with 4000 well-designed reactors, one would expect less than a 4% chance of a "Chernobyl-scale" reactor accident per century. If one estimates that such an accident might cause 20,000 eventual cancer deaths, the calculated risk over a century from reactor accidents would be of the order of 800 deaths. Reactors might do better or worse than this, but the anticipated scale of harm is in the ballpark of a thousand deaths per century – with large uncertainties in either direction.

For nuclear waste disposal, in a site such as Yucca Mountain, if the "maximally" exposed individual receives the proposed annual limit of 0.15 mSv, present estimates (based on the linearity hypothesis) suggest a 0.001% risk of an eventual fatal cancer. The maximum dose is reached only if the wastes are dissolved in a small volume of water, and therefore only a limited number of people would receive this dose. If this number were as high as 1000, the implied toll for Yucca Mountain neighbors would be one cancer fatality per century per repository site.[19] This toll would not start for many centuries, when the waste canisters begin to fail, and it not unreasonable to expect that cancer prevention and treatment will be much improved by then. Ignoring this prospect, and assuming many repositories and some doses above the prescribed limit, it still appears that the expected toll would be well under a thousand deaths per century.

It is much harder to gauge the scale of impacts for the unconfined risks. The most dramatic, and probably most immediate, of the dangers are those from nuclear weapons. Even a "small" nuclear weapon of the Hiroshima size could cause 100,000 deaths and a large-scale nuclear war could involve hundreds of millions of deaths. It is difficult to judge whether nuclear power would make the use of nuclear weapons more or less likely, or even

19The number of people who might be exposed is not well defined. The "critical group" considered in the National Academy study (NAS/NRC 1995) was expected to be less than 100.

whether, in the end, it would have a decisive effect in either direction. However, in view of the magnitude of the potential consequences, minimizing proliferation risks should be a central consideration in planning a nuclear energy future.

Turning to the other major areas, the picture is clearer, at least in terms of the direction of the impact. Nuclear power can help to reduce carbon dioxide emissions and thereby lessen the severity of predicted climate changes. It can also help ease the difficulties that will arise from the conflict between the shrinking of fossil fuel supplies and the rising material aspirations of a growing world population.

The full effects of global climate change are not well established. In addition to the expected increase in temperature as greenhouse gas levels rise, there will be an increase in sea level and very possibly increased occurrences of drought and violent climate events. Already such events are very costly in terms of human life. Tropical cyclones, hurricanes and typhoons caused approximately 500,000 deaths in the period from 1947-1980 and floods caused an additional 200,000 (Houghton, 1998, p. 3). Droughts in Africa were blamed for roughly 300,000 deaths in the 1980-89 decade. Clearly, if greenhouse gas effects exacerbate these problems, many lives will be lost throughout the world. In addition, there is the possibility of increased deaths from heat stress and the spread of insect-borne diseases (*ibid*, p. 132).

In another approach, economic estimates have been made of the effects of climate change. One summary of such estimates suggests that the worldwide economic impact of a 2.5 °C average rise in temperature would be over $500 billion per year (Fetter, 1999). The expenditure of $500 billion per year on nutrition, medical care, and accident prevention could obviously do a great deal for human health and survival, although presumably there would at best be only a partial transfer of resources. Nonetheless, it appears conservative to say that an annual saving of $500 billion, or anything approaching it, could lead to the saving of millions of lives per century.

Global warming costs reflect the adverse effects of using fossil fuels. There also can be adverse effects from having insufficient fossil fuels, especially oil and natural gas. In the absence of alternative energy sources, the shrinkage of fossil fuel resources combined with rising population is likely to exacerbate economic and political tensions, leading to increased chances of social upheaval and armed conflicts. The direct and indirect toll of even "low-level" conflicts results in many thousands of deaths per year, especially if the life-shortening effects of poverty are taken into account.

The effects of energy shortages are seen to be even greater if one approaches the matter in terms of ultimate world population. If energy shortages impose a limit upon the number of people that can be adequately supported on Earth, the impact is on the scale of billions of people.

It must again be emphasized that the numbers in the "quantitative" discussions of the preceding paragraphs provide nothing more than a crude hint as to the scale of the problems. But they are consistent with what is probably obvious without numbers: the risks from the open-ended problems are far greater than those from the confined problems. For the former, the potential worldwide toll is probably on the scale of millions of lives per century while for the latter it probably is on the scale of a thousand lives per century.

Steps to be taken

In summary, in order to address the critical problems of climate change and fossil fuel supply, we need greatly expanded sources of clean energy. It is dangerous to assume that non-nuclear sources will suffice. It is therefore important to strengthen the foundations upon which a nuclear expansion can be based, so that the expansion can proceed in an orderly manner – if and when it is seen to be needed – rather than with the haste of a crash program. Steps towards this end in the United States should include:

- Increased federal support for educational and research programs in nuclear energy, as already begun on a modest scale through the Nuclear Energy Research Initiative. In addition to the ideas and insights that individual projects provide, the existence of a federal program of adequate size would signal to students and the technical community in general that nuclear power has a credible and interesting future.
- Progress in the establishment of a permanent waste repository, presumably at Yucca Mountain, within a framework of reasonable waste disposal standards.
- Federal encouragement for the construction of prototypes of next-generation reactors, in the first instance for use in countries with large electricity markets, such as the United States, but for the longer run also with designs suitable for a broader array of countries. Without some government financial support or guarantees, it will be difficult to find commercial organizations willing to test whether American society – as reflected in actions taken by the courts, regulatory agencies, and local governments – will permit them to complete and operate new nuclear reactors.

A more limited, but potentially quite useful, step would be to establish in the United States a nuclear energy "think tank" where alternative nuclear futures could be analyzed critically by participants from universities, industry, government agencies, and private policy groups. The goal would be to further the safety and economy of nuclear power.

The contemplated nuclear revival would require a substantial change in U.S. federal policy. The private sector is unlikely to take the initiative in this direction. Ordering a new nuclear reactor offers little prospect of short-term gains, as long as the era of cheap natural gas continues, and would entail substantial financial risks if the construction were to be delayed by public opposition or regulatory difficulties. For a nuclear revival to occur in the near future government leadership will be important, and this in turn will require a new climate of public opinion. The most promising agent for fostering a change in public attitudes would be a new group of environmental revisionists, who conclude that – when taken all in all – the dangers of trying to do without nuclear power are of greater scope and potential severity than the dangers created by using it.

REFERENCES

Belles, R.J. *et al.*, 1998, *Precursors to Potential Severe Core Damage Accidents: 1997, A Status Report*, NUREG/CR-4674, Vol. 26, U.S. Nuclear Regulatory Commission, Washington, D.C.

Bodansky, David, 1996, *Nuclear Energy: Principles, Practices and Prospects*, American Institute of Physics Press/Springer-Verlag, Woodbury, N.Y.

Bodansky, David, 2000, "Nuclear Energy and the Large Environment," *Physics and Society* 29, no. 1:4.

Breidenbach, L., 1997, "Thermo-economic evaluation of a nuclear co-production plant for electricity and potable water," in *Nuclear Desalination of Sea Water*, Proceedings of an International Symposium on Desalination of Seawater with Nuclear Energy (Taejon, South Korea), IAEA, Vienna.

Cohen, Joel E., 1995, *How Many People Can the Earth Support?*, W.W. Norton & Co., New York.

Daily, G.C., Ehrlich, A.H., and Ehrlich, P.R., 1994, "Optimum Human Population Size," *Population and Environment, A Journal of Interdisciplinary Studies* 15:469.

DOE/EIA 1998, *Emissions of Greenhouse Gases in the United States 1997*, Report DOE/EIA-0573(97), U.S. Department of Energy, Washington, D.C.

DOE/EIA 1999a, *Annual Energy Review 1998*, Report DOE/EIA-0384(98), U.S. Department of Energy, Washington, D.C.

DOE/EIA 1999b, *Emissions of Greenhouse Gases in the United States 1998*, Report DOE/EIA-0573(98),

U.S. Department of Energy, Washington, D.C.
EPA, 1985, "40 CFR Part 191, Environmental Standards for the Management and Disposal of Spent Nuclear Fuel, High-Level and Transuranic Radioactive Wastes; Final Rule," *Federal Register* 50, no. 182:38066.
EPA, 1999, "40 CFR Part 197, Environmental Radiation Protection Standards for Yucca Mountain, Nevada; Proposed Rule," *Federal Register* 64, no. 166:46976.
Fetter, Steve, 1999, "Preventing climate change: the role of nuclear energy," in *Nuclear Energy, Promise or Peril?*, B.C.C. van der Zwann, ed., World Scientific, Singapore.
GE 1999, *The ABWR Plant General Description* , GE Nuclear Energy, Palo Alto, Ch. 10.
Gleick, Peter H., ed., 1993, *Water in Crisis, A Guide to the World's Fresh Water Resources*, Oxford University Press, New York.
Hardin, Garrett 1996, *Population and Environment, A Journal of Interdisciplinary Studies* 18:73.
Houghton, John, 1997, *Global Warming: The Complete Briefing,* Second Edition, Cambridge University Press, Cambridge.
IAEA, 1996, "Summary of the Conference Results," in *One Decade After Chernobyl, Summing up the Consequences of the Accident,* International Atomic Energy Agency, Vienna.
IAEA, 1998, *Nuclear Power Reactors in the World,* April 1998 Edition, Reference Data Series No. 2, International Atomic Energy Agency, Vienna.
IPCC, 1995. *IPCC Second Assessment: Climate Change 1995, A Report of the Intergovernmental Panel on Climate Change*, World Meteorological Organization and UN Environment Programme.
Kerr, Richard A, 1998, "The next oil crisis looms large – and perhaps close," *Science* 281:1128.
Kupitz, Juergen, 1995, "Nuclear energy for seawater desalination: updating the record," *IAEA Bulletin* 37, no. 2: 21.
Murley, T.E., 1990, "Developments in Nuclear Safety," *Nuclear Safety* 31, no. 1: 1.
NAS/NRC 1995, *Technical Bases for Yucca Mountain Standards*, Committee on Technical Bases for Yucca Mountain Standards, National Research Council, Robert W. Fri, ch., National Academy Press, Washington, D.C.
OCWRM, 1998. *Viability Assessment of a Repository at Yucca Mountain, Volume 3: Total System Performance Assessment,* DOE/RW-0508/V3, U.S. Department of Energy, Office of Civilian Radioactive Waste Management, North Las Vegas.
Pimentel, D. *et al.*, 1994, "Natural resources and optimum human population," *Population and Environment, A Journal of Interdisciplinary Studies* 15:347.
Travers, William D., 1999, letter of November 2, 1999 to S. D. Page, Director of Office of Radiation and Indoor Air, EPA. [http://www.nrc.gov/OPA/reports/epa1199.htm]

NUCLEAR ENERGY AND SECURITY

Thomas E. Blejwas, Thomas L. Sanders, Robert J. Eagan, and Arnold B. Baker

Sandia National Laboratories*
P.O. Box 5800
Albuquerque, NM 87185

INTRODUCTION

Nuclear power is an important and, we believe, essential component of a secure nuclear future. Although nuclear fuel cycles create materials that have some potential for use in nuclear weapons, with appropriate fuel cycles, nuclear power could reduce rather than increase real proliferation risk worldwide. Future fuel cycles could be designed to avoid plutonium production, generate minimal amounts of plutonium in proliferation-resistant amounts or configurations, and/or transparently and efficiently consume plutonium already created. Furthermore, a strong and viable U.S. nuclear infrastructure, of which nuclear power is a large element, is essential if the U.S. is to maintain a leadership or even participatory role in defining the global nuclear infrastructure and controlling the proliferation of nuclear weapons.

By focusing on new fuel cycles and new reactor technologies, it is possible to advantageously burn and reduce nuclear materials that could be used for nuclear weapons rather than increase and/or dispose of these materials. Thus, we suggest that planners for a secure nuclear future use technology to design an "ideal" future. In this future, nuclear power creates large amounts of virtually atmospherically clean energy while significantly lowering the threat of proliferation through the thoughtful use, physical security, and agreed-upon transparency of nuclear materials. We must develop options for policy makers that bring us as close as practical to this ideal. Just as "Atoms for Peace" became the ideal for the first nuclear century, we see a potential nuclear future that contributes significantly to "power for peace and prosperity."

THE NEED FOR NUCLEAR POWER

Most of the arguments for nuclear power are well known to participants in this conference. Nuclear power does not generate carbon dioxide as a part of the fuel cycle

*Sandia is a multiprogram laboratory operated by Sandia Corporation, a Lockheed Martin Company, for the U.S. Department of Energy under Contract DE-AC04-94AL85000.

(except for small amounts associated with mining and other operations); it uses a highly concentrated fuel that is relatively abundant in the U.S. (and with breeder reactors could supply power for centuries); and it has an outstanding safety record in the U.S. Numerous authors have made compelling arguments for the need for nuclear power in our future (see, for example, America the Powerless, Facing our Nuclear Energy Dilemma by Allen D. Waltar, which is both a very complete and very readable treatise).

In addition to reasoned analyses in articles and books that support the need for nuclear power, some analysts have created models and have used those models to show potential impacts and benefits from nuclear power by examining different scenarios (see, for example, Krakowski). These models are generally the basis for debate and discussion among "insiders" in the nuclear arena, but often are not exposed to or applied by decision-makers.

Recently at Sandia National Laboratories, teams led by Dr. Arnold Baker have developed a series of reduced form dynamic simulation models that integrate energy, economic and environmental aspects of complex systems, run on lap-top PCs, and allow the user to easily vary input parameters. One model, the USEGM, (a U.S. energy and greenhouse gas model developed by Arnold Baker, Thomas Drennen, Orman Paananen and David Harris) integrates U.S. energy markets and carbon emissions by energy use sector and fuel through 2020. It is a demand side model that is largely driven by income, prices and energy efficiency.

The first curve in Figure 1* presents a base case from that model for U.S. carbon emissions. This case is benchmarked to the energy use and mix patterns in the DOE/EIA 1999 Annual Energy Outlook Reference Case through the year 2020. Note that emissions increase 48% from 1990 to 2020, while the goal of the Kyoto Protocol is a 7% decrease by 2008-2012. The second curve in Figure 1 was generated by assuming that nuclear energy and renewables grow to 50% of electricity production in 2020. This reduces carbon emissions in 2020 by 13% from the base case, but emissions would still grow substantially.

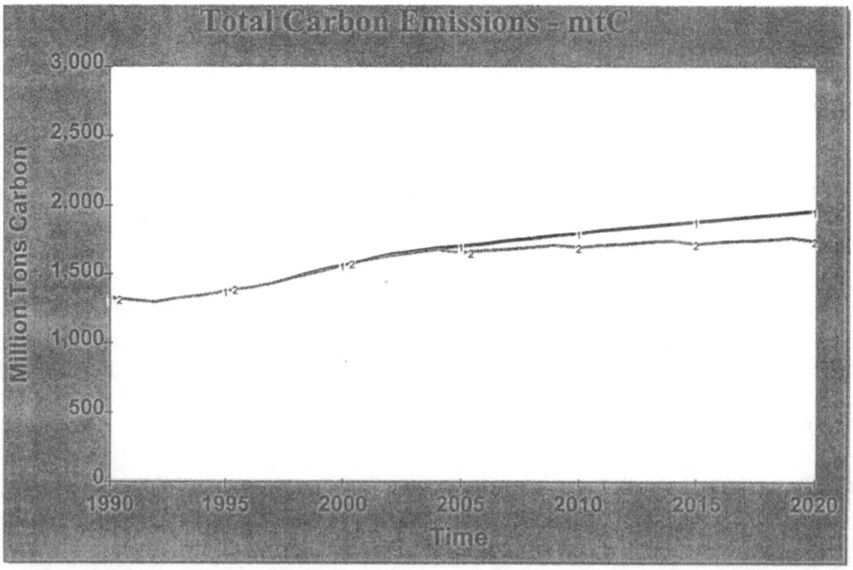

Figure 1. Base case (1) compared to scenario (2) with nuclear and renewables increased to 50% by 2020.

*Figures are actual screen prints from the PC model.

CURBING CARBON EMISSIONS

What would it take to significantly curb carbon emissions, while allowing continuing increases in energy consumption? In Figure 2, we assume again that the non-carbon electric share is increased to 50% (as in Figure 1) and, additionally, that 20% of transportation vehicles are electric by 2020. Obviously, the improvements in emissions are small, but in the right direction.

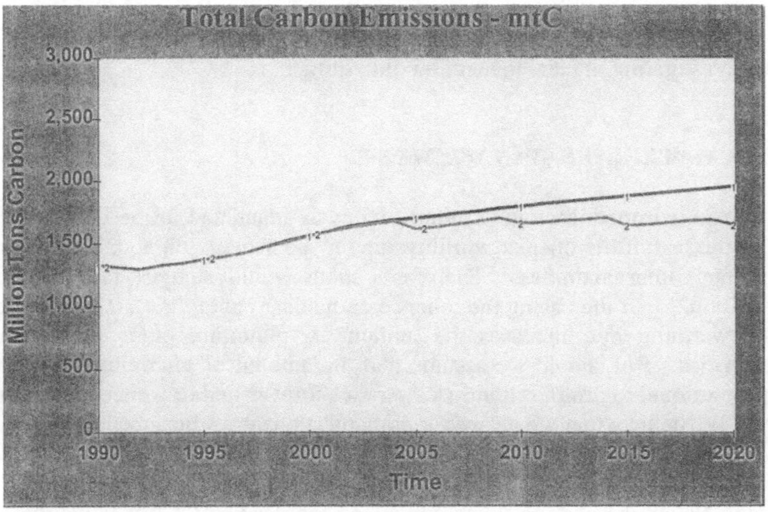

Figure 2. Scenario (2) with 20% electric vehicles and 50% non-carbon-based electricity by 2020 compared to base case (1).

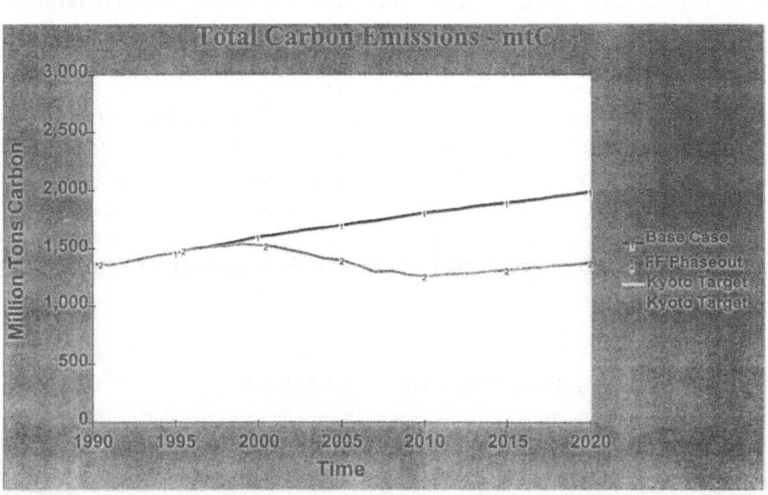

Figure 3. Scenario (2) with 100% of electricity generated by nuclear and renewables compared to base case (1) and Kyoto target.

Based on our experience with the models, measures more drastic than can be imagined in twenty years are necessary. For example, in Figure 3, we show the results of replacing all coal, oil, and natural gas-fired generation of electricity with renewables and nuclear energy by 2010. Such draconian measures would significantly reduce carbon emissions, but are clearly unrealistic. And even in this case, U.S. carbon emissions rise between 2010 and 2020 above the Kyoto target in response to U.S. economic growth.

To permit a fuller exploration of such issues, we are developing models that look further into the future and allow more complex relationships to be included. We expect these models to show that significant changes in transportation and manufacturing systems will be necessary, in addition to changes in the production of electricity. It is worth noting that rather than converting vehicles to battery-supplied electricity, we could have assumed that transportation systems convert to hydrogen as a fuel (e.g., hydrogen fuel cells) and the hydrogen is generated with nuclear power; but the model cannot presently accommodate this scenario. A Nuclear Energy Research Initiative (NERI) activity at Sandia National Laboratories, funded by the U. S. Department of Energy (DOE), is investigating nuclear generation of hydrogen.

NUCLEAR MATERIALS MANAGEMENT

The so-called once-through uranium fuel cycle mandated in the U.S. and common in much of the rest of the nuclear world results in spent fuel with significant amounts of plutonium and other actinides. Therefore, some would suggest (see, for example, Krakowski, et al.) that increasing the reliance on nuclear energy may reduce the potential for global warming but increases the amount of plutonium and, consequently, the proliferation risk. But should we assume that the amount of plutonium in spent fuel is directly proportional to proliferation risk? Are we forever tied to a once-through uranium fuel cycle? We believe that the answer to both questions must be a resounding no!

Unfortunately, our models cannot presently accommodate alternative fuel cycles, recycling, or other new technologies; but the possibilities are numerous. For example, Los Alamos National Laboratory has advocated the accelerator transmutation of waste (ATW) and Congress recently allocated $9M for research on advanced spent-fuel treatment, with emphasis on the ATW. General Atomic advocates the burning of plutonium using a graphite-moderated, gas-cooled, thermal-neutron reactor in conjunction with an accelerator. There is considerable attention, internationally, on the use of fast-neutron reactors for converting fertile material to fissile material, generating energy, and burning actinides. The burning of mixed oxide fuels (MOX) in conventional light-water reactors is one approach to consuming weapons plutonium, with the residual plutonium in hot spent fuel. Other schemes for more directly burning plutonium have been proposed and should be investigated further. Many have suggested a thorium fuel cycle that would take advantage of the worlds large thorium resources and, potentially, provide a fuel cycle that is more proliferation resistant than the once-through uranium cycle. The possibilities for the future are exciting. If we can develop clear paths forward, successfully inform policy makers, and perform appropriate research; the future can be more secure.

GLOBAL NUCLEAR MATERIALS MANAGEMENT

Global nuclear material management, started at Sandia National Laboratories as a visionary concept for tying the national security benefits of materials back to proliferation prevention, arms control and civilian nuclear power. Under the leadership of Senator

Sam Nunn, the Center for Strategic and International Studies (CSIS) organized two workshops on GNMM. To quote results of the CSIS GNMM Policy Forum, July 1999:

"The vision of Global Nuclear Material Management (GNMM) is of a world in which all nuclear materials are safe, secure, and accounted for, from "cradle-to-grave,"with sufficient transparency to assure the world that this is the case. That is a daunting goal, which must be approached step by step, within well-defined strategic framework. This panel has identified two key areas where the need for action is particularly urgent:

> **Eroding controls in the former Soviet Union.** Insecure and oversized nuclear weapons and materials stockpiles in the former Soviet Union, with little transparency in their management, coupled with an oversized and underfunded nuclear complex, pose severe threats to U.S. and international security. The possibility that the essential ingredients of nuclear weapons could fall into the hands of terrorists and proliferating states is all too real, and immediate actions are needed to reduce this threat to the security of America and the world. . . .
>
> **A withering foundation for U.S. leadership.** Judged by any of a broad range of criteria, the infrastructure of U.S. leadership in nuclear technologies has greatly weakened over the last two decades. U.S. nuclear Research & Development (R&D) is dwarfed by R&D underway in other nations, the cadre of experienced personnel is dwindling, and nuclear engineering departments at U.S. universities are shrinking. The United States has virtually disengaged from international discussions and cooperation on the future of the nuclear fuel cycle. If the United States can no longer credibly claim a leadership role in nuclear technology, or is seen as having no interest in the future of nuclear energy, its ability to lead in nonproliferation could be substantially undermined."

Furthermore, the GNMM report recommends that "immediate action be taken to rebuild the R&D program, the cadre of experts, the R&D facilities, and material infrastructure that help provide the foundation for global leadership." Many of us at the DOE's weapons laboratories see GNMM as a companion effort to stockpile stewardship. Clearly, the future of nuclear energy must be integrated with the global management of nuclear materials.

PUBLIC ATTITUDES TOWARDS THINGS NUCLEAR

We have all heard that public attitudes will prevent a future reliance on new nuclear power systems. However, independent polling by Hank Jenkins-Smith at the University of New Mexico (some of which was sponsored by Sandia National Laboratories), found a somewhat different picture:

"First, Americans do not want to abandon nuclear energy. When a nationwide sample of Americans were asked whether the current utilization of nuclear energy in the United States should be decreased, kept the same, or increased, about 43% wanted to keep it the same and around 30% wanted to increase it. Approximately 27% wanted to decrease reliance on nuclear energy.

Second, most Americans would like the government to investigate prospects for reusing spent nuclear fuel rods, even when apprised of the possible proliferation risks associated with reprocessing. In fact, whether it is called "reusing" or "recycling" spent nuclear fuel, about 4 out of 5 respondents to a random sample of Americans were in favor of making use of spent fuel to produce more energy. . . .

The point behind these examples is that Americans do see substantial benefits in the use of nuclear technologies, whether they be for energy or national security. But these benefits are not addressed in our fragmented nuclear policy discussion concerning nuclear waste management. When it comes to waste, regardless of who asks, most Americans are opposed to having waste shipped through their communities or disposed of in facilities in their states. Why is that?

A lot of our research has been focused on why people react as they do to the prospect of nuclear waste transport and storage. In a nutshell, when faced with a controversial problem like nuclear waste, Americans want to hear good and robust reasons for a policy. They want to see that the solution offered is a long term one. And they want to be able to identify tangible national benefits from the policy." We believe that an "ideal" nuclear future could have sufficient tangible national benefits for the American public to react positively.

TOWARDS AN "IDEAL" NUCLEAR FUTURE

Our view of an ideal nuclear future is one in which nuclear is well positioned to be a substantial contributor to the concept of "power for peace and prosperity." It includes the following: First, nuclear energy would be fully cost-competitive and plentiful and would contribute significantly to the avoidance of carbon emissions in our atmosphere. Second, through a combination of advanced nuclear fuel cycles and nuclear technologies, amounts of fissile materials largely would be reduced to those necessary for energy production and limited nuclear weapons use. Nuclear weapons would not exist beyond the existing weapons states. Third, any fissile materials, whether separated or in spent fuel, would be safe, physically secure and transparent through the implementation of international agreements and participatory R&D.

Reaching this future nuclear state is a daunting task. What is required? First, we have to create a vision of a sound, integrated, pragmatic nuclear future. Hopefully this paper will contribute to that vision in some small way. Technical approaches for achieving the vision must be developed. We believe that models that are extensions of the dynamic simulation models presented above could help define the range of possibilities. In the hands of experienced planners, such models could help define where advanced technologies could have the greatest impacts. By using such models to interact with policy-makers and their technical staff, more informed decisions about R&D funding can be made. With the results of sound research and further advanced models, we will have prepared our future administrations and congresses to negotiate internationally and to put in place comprehensive energy, nuclear, and national security policies.

A LEVEL PLAYING FIELD

Creating an economic and/or environmental level playing field for nuclear power may be possible, but the aspects that nuclear shares with nuclear weapons can never be level with other fuel sources. A common misconception is that eliminating nuclear energy would help our proliferation problems by eliminating the generation of tons of plutonium-containing spent fuel. But as a practical matter, if the U.S. abandoned nuclear energy, the use of nuclear energy outside the U.S. still would continue, and the U.S. would weaken seriously its ability to deal with proliferation issues. The proliferation threat of U.S. spent nuclear fuel is insignificant compared to the real risk of loss of control of separated fissile materials in the former Soviet Union, for example. As noted in the GNMM report, another real threat is the loss of nuclear infrastructure and any kind of leadership position for things nuclear. Also, as noted above, future nuclear fuel cycles give the U.S. the potential to burn or otherwise reduce fissile materials. Therefore, we must find ways to help U.S. policy makers support development of an integrated U.S. nuclear policy, despite the complexity of the issue and the complexity of our political system. Such an integrated politically acceptable policy is the only way to achieve the potential energy, economic and environmental benefits from nuclear power and the protection from nuclear weapons and materials that the world demands.

ACKNOWLEDGMENT

"Sandia is a multiporgram laboratory operated by Sandia Corporation, a Lockheed Martin Company, for the United States Department of Energy under Contract DE-AC04-94AL85000."

REFERENCES

A.D. Waltar, America the Powerless, Facing Our Nuclear Energy Dilemma, Cogito Books, Madison, Wisconsin, 1995.

R.A. Krakowski, L. Bennett, and E. Bertel, "Nuclear Fission: For Safe, Globally Sustainable, Proliferation-Resistant, and Cost-Effective Energy," Proceedings of the International Conference on Preparing the Ground for Renewal of Nuclear Power, held October 22-23,1998, edited by B.N. Kursunoglu, et al., Kluwer Academic / Plenum Publishers, 1999.

Center for Strategic and International Studies, Global Nuclear Materials Management Policy Forum, July 1999, Sam Nunn, Chair.

H. Jenkins-Smith, Congressional Testimony, May 1998.

Energy Problems of The Future
Can We Solve Them?

Bertram Wolfe
Monte Sereno, CA. USA

1. Introduction

We are in a changing world. It has taken two thousand years for the population of the world to grow from 250 million to the near 6 billion people today. But in the next fifty years world population is projected to grow to about 10 billion people.[1] How will future world needs be met?

The primary feature of the projected population growth is the increase of the third world from 41/2 to 8 billion people in the next half century.[1] History has shown that the key measure of population welfare and population stabilization is energy use. Today, people in the high birthrate, poverty stricken, low income nations use only a small fraction of the per-capita energy use of the rest of the world.[2]

Suppose that in the next 50 years a massively successful world conservation program leads to acceptable living standards, and stabilization of world population with a per-capita energy use only a third of today's U.S. use. Then world energy needs will triple.[2] How can such energy needs be met?

There are several key problems. One, is the long term availability of fossil fuels which today supply some eighty per cent of world energy.

On our present course it is questionable whether economic oil and gas supplies will be available by the latter part of the century[3]. Although coal may be available for a century or two thereafter, it is more difficult and costly to transport throughout the world. Energy shortages may prevent the stabilization of world population at a decent standard of living.

In addition there is the possibility of major international hostilities over scarce energy supplies. Ask yourself why the U.N. and the U.S. fought Iraq in Kuwait; or why the U.S. still maintains troops in Saudi Arabia - despite the deaths by terrorists of a number of them.

The problem of most vital concern today, the subject of the international conference last year in Kyoto, Japan, is that of possible disastrous earth warming due to the increasing atmospheric emissions from fossil fuel burning. This problem should be addressed early since the CO_2 emitted by fossil fuels now will remain in the atmosphere for many decades. Unfortunately, no global solution was identified at Kyoto.

Although the great majority of atmospheric scientists believe that the predictions of earth warming from fossil fuel burning are valid, there are some that express doubts about the predictions. The National Academy of Sciences indicated in their 1990 book, "One Earth One Future"[4] that there were indeed uncertainties in the difficult calculations; but they noted that if CO_2 emissions continued to increase, future earth warming may be greater, rather than less than projected. Thus, it does not seem responsible to wait for proof, when it may then not be possible to correct the situation.

Finally, one might note that there are concerns over problems of radioactive wastes tens of thousands of years in the future. Should there not be even more concern over the lack of needed fossil fuel supplies only a hundred or two hundred years from now. Fossil fuels are vitally needed for special energy tasks and particularly, for special non energy uses such as chemical and manufacturing production.

2. Is There an Energy Solution for the Future?

There is only one practical, solution to the pending world energy problems: Nuclear Energy. Maybe economic large scale solar or wind power, or fusion or cold fusion, or energy from satellites in space, or some other massive new clean energy source can be developed. We should keep working on them; but the practicality of large scale energy production from such sources is so questionable that it would be irresponsible to count on them. Consider, for example, that a solar plant with the same electricity output of a large nuclear plant would require an area of 50 to 100 square miles. Its costs and problems would be prohibitive even if it could be constructed for the same square foot cost as a highway sign; and because of its size and the daily and yearly changes in the incoming solar heat, its operation and maintenance costs may also be prohibitive. And consider the environmental affects of the 50 to 100 thousand square miles of solar panels, if today's U.S. electrical supply was converted to solar power. Similar problems exist for the other sources under development.

The only energy source available to significantly ameliorate the coming energy crisis is nuclear energy. Nuclear Energy emits no global warming gases. Nuclear energy today supplies some 7% of the world's energy. It supplies 17% of the world's electricity - more than the total electrical energy supplied at the start of the peaceful nuclear era in 1954. Nuclear energy plants, built and operated to U.S. standards around the world, have not harmed any member of the public. (Chernobyl would not have been allowed to be built in the west, or operated as it was.) The safety, economics, and practicality of nuclear energy have been demonstrated over the past few decades and there are now some four hundred nuclear plants operating around the world.

Nuclear energy could meet the energy needs of the word almost indefinitely. Assume that to meet world needs by the middle of the next century, nuclear energy is targeted to provide half the needed energy; that is, one and a half times today's yearly world energy use. This would require the construction of about a hundred new modern nuclear plants a year for the next fifty years. Can this be done?

It would not be a simple task, but one might note that in the late sixties and early seventies some thirty to forty nuclear plants were being ordered in the U.S. and projections were that there would be a thousand nuclear plants on the line in the U.S. by the end of the century.

Because of the Arab oil embargo of 1973 and the subsequent increase in energy costs, U.S. energy growth decreased and U.S. nuclear capacity has increased from forty to only a hundred and ten plants. The point is that although the need disappeared, there was little doubt that tens of nuclear plants per year could be installed in just the U.S.; and there is little doubt that world wide, the construction of a hundred new plants per year can be accomplished.

3. Implementing the Nuclear Solution to the World's Problems

In principle, the energy and nuclear energy needs of the world could be met individually by each nation. But to meet the worldwide installation needs of 5000 new nuclear plants efficiently and safely, a world program would be helpful, if not vital.

An international nuclear safety organization could be set up to adopt world safety standards, and approve standardized plant designs which could be built efficiently in volume around the world. This international organization, or additional ones, could approve, or provide, standard operator and maintenance training; and like the IAEA could provide periodic inspections to assure that safety standards are being maintained, and that nuclear materials were not being illegally diverted. Indeed, perhaps the IAEA could be expanded to meet these requirements.

In addition it may be helpful to set up, or at least approve, a number of world specialized manufacturing facilities; for example, quality facilities to manufacture the hundred pressure vessels needed each year. Laboratories should be available to expeditiously solve reactor operating problems which may arise worldwide, and to improve reactor performance. And there should be an international program to develop the Fast Reactor which generates some sixty to a hundred times as much energy from a pound of uranium than do our present commercial reactors.

With the expansion of nuclear power, the Fast Reactor is likely to be needed in the next half century as supplies of uranium for the present reactor types become scarce. Because of its efficiency in utilizing uranium the Fast Reactor can supply the world's energy needs indefinitely.

A characteristic of nuclear energy development, which is evident from past experience, is that it will likely take several decades to find and resolve the problems which could impede the reliable operation of a new type of nuclear energy plant. Consider the past problems and note that we are still having problems with steam generators in our operating plants. Similarly, the Japanese Monju and the French Superphoenix fast reactor plants encountered problems which are delaying future development. Although the fast reactor will not be needed for at least several decades, it is clear that if we wish to responsibly prepare for the future we should be vigorously pursuing its long term development now.

Perhaps there should be centralized areas for storage of spent fuel from present reactor types. These storage areas should be adjacent to reprocessing plants which would be built to process the spent fuel to provide the new fuel for the Fast Reactors going on line.

The fast reactor high level waste is accumulated at the reprocessing plants and retains its toxicity for only a few hundred years, rather than the tens of thousands of years of the spent fuel wastes from our present reactors,. Thus, the nuclear waste disposal problems are minimal and arrangements for disposal could be made on a global basis.

Finally, it is likely to require international financial programs in order to provide the energy and distribution facilities, and supplies, needed by the growing low income nations.

Nuclear plants (as well as fossil fueled plants) are today being added individually in third world countries such as China, India, and Korea, and in growing industrialized countries such as Japan. But the point of the above discussion is that the increasing energy use and needs in the coming decades will have global effects which will effect the total world population. To meet the immense world requirements in a timely, safe and environmentally sound manner may require the problem to be approached globally, rather than individually by nations needing new energy supplies.

4. Nuclear Energy Problems

There are two interconnected problems which could prevent Nuclear Energy from expanding to meet coming world needs.

The first is an incorrect public perception about the risks of nuclear energy relative to its benefits. The nuclear industry has done a poor job in educating the public to the fact that not a single member of the public has been harmed by nuclear plants which were built and operated to U.S. and western world standards. Indeed, even at Chernobyl, which would not have been allowed in the western world, the great majority of deaths were not due to nuclear effects. There were some 40 operating personnel who died working to contain the accident, and in addition several hundred children contracted thyroid cancers due to drinking contaminated milk, and four of them have died. But the major cause of Chernobyl deaths was the abortion of some eighty thousand babies by mothers who were frightened by the radiation spread over Europe. They were not told that the radiation was small compared to their normal radiation exposure from nature (which in fact, may be healthful).

Similarly, members of the public are concerned about nuclear wastes, which are very small in volume, have not harmed anyone, and have risks very small relative to the potential dangers of increasing fossil fuel use.

The point is that there are no technical nuclear problems without rational solutions. The key problems are public misconceptions and the resulting institutional impediments which can impact on the potential of nuclear energy to meet future world needs.

5. Consider the situation in the United States:

Today , the U.S. is the major world energy user. It uses a quarter of the total energy used in the world and emits about a quarter of the CO_2.[2] It has nuclear energy problems which are institutional, not technical; and these could imperil the solution to the coming world's energy situation.

The development of nuclear energy has been a recognized accomplishment of the United States. The U.S. has been the leader in the initial development of safe, efficient nuclear power plants, and indeed, almost all the nuclear plants in the world are based on the nuclear plants originally developed in the U.S., with technology transferred abroad. Advanced U.S. designed plants are being built today in such countries as Japan, South Korea, and Taiwan. They may soon be built in China, when the U.S./China agreement permitting the transfer of U.S. nuclear technology to China is approved by the U.S. congress.

But nuclear energy has problems in the U.S.. Since the 1973 Arab oil boycott there has been a surplus of electrical capacity in the U.S.; and no large base load electrical plants have been ordered. Indeed, some 100 nuclear plants and some 80 coal plants on order were canceled after 1973. The sixty nuclear plants put on the line since 1973 (providing 40% of new electricity capacity) were all ordered before 1973. And because there was no urgent need, bureaucratic licensing procedures and litigious court attacks by anti-nuclear groups have led to construction times of a dozen to twenty years, and uneconomic costs. This compares to the four to six year construction times of the U.S. reactors built abroad; and indeed to the four to six year construction times in the U.S. prior to 1973.

In the last twenty odd years, almost all nuclear endeavors in the U.S. have run into bureaucratic and litigious delays, making their schedules and costs unpredictable. In addition to reactor construction, there are the bureaucratic delays in the waste repository programs. Another example is the attempt to build a new uranium enrichment plant in the state of Louisiana, a plant which uses advanced technology demonstrated in several countries in Europe. Licensing started over seven years ago and is still held up by issues without relevance to technology or safety. It is approaching the point where the delays, and costs may lead to the abandonment of a potential asset.

Because of the uncertainties, costs, and political problems there is not a single U.S. utility willing to take the risk of building a new nuclear plant in the U.S., despite the favorable experience abroad. Indeed, no utility was even willing to help the Nuclear Regulatory Commission (NRC) test its new site qualification procedure, even though it meant no firm commitment to proceed with an actual nuclear construction project. The situation is so bad that the Energy Information Agency (EIA) projects the shutdown of 40% of present U.S. nuclear capacity in the next 20 years, without a single new plant being built.

At present there is still excess electrical capacity in the U.S.. In addition, the low price of gas permits the economic construction of new gas turbine electrical plants. But new gas plants, and methane leaking from underground gas production facilities and pipelines, exacerbates the global warming problem. In addition there is the question of the future availability and price of gas.

Even if the price of gas were to double, or a large CO_2 tax was imposed to discourage greenhouse gas emissions, nuclear power would still be in trouble in the U.S.. Despite new licensing procedures, it is doubtful that a rational businessman would commit to a new billion dollar project without assurance that, with the new licensing system, unnecessary delays and uneconomic costs would be eliminated. The U.S. government, which has taken a strong position on the need to reduce CO_2 emissions, should not delay in solving this problem, but should initiate and take the financial risks on several new projects that would demonstrate that U.S. designed nuclear plants can be built as efficiently in the U.S. as they are abroad. These early demonstration plants could take the place of those non CO_2 emitting nuclear plants being shut down because they are reaching the end of their lifetimes. And they would provide the experience needed to allow the U.S. to proceed with an expansion of nuclear power when national needs for CO_2 emission reductions are implemented.

Another important reason to proceed with a demonstration nuclear expansion is to help the U.S. maintain its leadership and influence abroad. As indicated above, the U.S. has been a leader in world nuclear energy development and implementation. In view of the need for a major world expansion of nuclear energy it would be unfortunate if the U.S. was unable to provide the needed help to the world.

Because of its unnecessary institutional impediments to the expansion of nuclear energy, the U.S. may be unable to meet its goals of reducing Greenhouse gas emissions, and can lose its ability to lead in solving the global energy problem. Further, because of its world economic position, and its large energy utilization the U.S. could be a major impediment to the solution of the world's energy problems. Thus, for the benefit of the world and its own people, the non technical, institutional impediments in the U.S. should be rapidly removed.

Such actions should be taken in other countries such as Italy, Sweden, and Germany which also institutionally prevent the growth of nuclear energy.

5. Conclusions:

In the coming decades, due to an expanding world population and an increase in world living standards, this planet faces an ever increasing need for energy. Meeting the increasing needs with fossil fuels, which today supplies 80% of world energy, may lead to fuel shortages, disastrous international hostilities over limited supplies, and to calamitous environmental effects due to increasing CO_2 emissions.

Perhaps immense new economic fossil fuel supplies will be unearthed; and perhaps it will be found that the projected warming of the earth from fossil fuel gas emissions does not take place. On the other hand, the problems may be much more intense than projected, and continuing on our present fossil fuel course will make the problems more difficult, if not impossible, to rectify in the future.

The one available means to significantly meet future world energy needs, while avoiding the fossil fuel problems, is to expand the use of nuclear energy. The key problems which could impede this nuclear energy expansion are not technical. The problems are unnecessary institutional impediments present in some nations, and the need of a global plan to provide for the needed expansion.

Both of these problems can be resolved, and should be, on an expeditious basis. This could save mankind from unnecessary future calamities.

References

1. World Population Projection, 1994-5 Edition — The World Bank
2. 1997 World Development Indicators — The World Bank
3. Energy For Tomorrow's World — World Energy Council, 1993
4. One Earth One Future — National Academy of Sciences, 1990

PUBLIC AND POLITICAL SUPPORT FOR NUCLEAR ENERGY

Scott Peterson

Senior Director, External Communications
Nuclear Energy Institute
Washington, DC 20006

GLOBAL FOUNDATION

I'm pleased to be here to give you some good news about the rediscovery of nuclear energy. If you read the Wall Street Journal October 19, you know that predictions of the demise of nuclear energy have been greatly exaggerated. Indeed, such predictions have been completely off the mark.

To quote the Journal: "Not since the days of bell-bottoms, disco and oil embargoes have so many big companies been so concerned with energy. Nuclear power plants, long the *bete noire* of environmentalists, this time are getting off scot-free. Nukes don't emit the greenhouse gases believed responsible for global warming."

As we move into a new era in the business of providing electricity, nuclear energy is looking more and more attractive. And public and political support for nuclear energy is growing. Underlying this public and political support is the growing recognition of the key role nuclear energy plays in protecting the environment.

A NUCLEAR RENAISSANCE AND A NEW ATTITUDE

Fundamental to the support for nuclear energy is the new attitude in the industry itself. Nuclear electricity producers are adapting well to the onset of competition in the marketplace. We are seeing consolidation, nuclear plant purchases, and the pursuit of license renewals. In this new environment, the U.S. nuclear industry is entertaining something of a renaissance and beginning to take credit for its accomplishments.

Nuclear plants operate in 17 of the 24 states that have opened their electricity markets. The nuclear facilities in those 17 states account for 60 of the 103 operating reactors in the United States.

The Challenges to Nuclear Power in the Twenty-First Century
Edited by Kursunoglu *et al.*, Kluwer Academic/Plenum Publishers, New York, 2000

Two companies are in the process of renewing their operating licenses at five reactors with the Nuclear Regulatory Commission for an additional 20 years of operation. We expect the NRC to approve the first of those requests—Baltimore Gas and Electric's application for its two-unit Calvert Cliffs Nuclear Power Plant—next March.

Moreover, the owners of 24 other generating units have notified the NRC that they intend to pursue license renewal as well. And at least 10 other units have indicated informally an interest in doing so.

In short, we expect that most nuclear power reactors will be renewing their licenses eventually. Clearly, owners of these nuclear units believe that operating for an additional 20 years is a good business decision.

Equally important, a number of nuclear plants are being purchased by companies seeking to expand their nuclear holdings, with other sales in the works.

Most recently, PECO Energy and Commonwealth Edison's parent company Unicom announced plans to merge. Once finalized, the nearly $32 billion deal will mean that the new company will own and operate 14 nuclear units...not counting AmerGen—PECO Energy's joint venture with British Energy—that is in the process of purchasing six additional nuclear units so far.

What are the reasons for this confidence? There is a growing realization that a nuclear power plant—operated safely and efficiently—is an attractive investment for companies. And, contrary to conventional wisdom, nuclear power plants are even more attractive in a competitive market.

Nuclear output from January to June 1999 was up 9.5 percent over the first six months of last year—347 billion kilowatt-hours compared with 317 billion kilowatt-hours in 1998. That's about 15 percent more electricity from nuclear power that during the first six months of 1997.

U.S. nuclear plants are on track to set a new record for annual production. In 1998, the industry had a record capacity factor of 79.5 percent. Moreover, 43 plants operated at over 90 percent capacity, and 32 operated at over 80 percent capacity. The capacity factor for all U.S. plants through August of this year is 85 percent.

As the industry improves on this outstanding production and reliability record, it continues to improve its safety record as well. For example, in 1998 the average number of significant events per unit—an important indicator of safe operation—improved to point zero four (0.04), a sharp decrease from 2.5 in 1985.

The industry also accomplished these feats while keeping production costs competitive with coal, and well below those of other fuels, including new natural gas plants.

Another important reason for this new attitude is the remarkable progress being made in changing the regulatory process. The excellent performance of our plants, the experience gained through more than 2,200 reactor years of operation, the availability of important tools like probabilistic safety assessments, and the competitive electricity market have fostered a reassessment of how the Nuclear Regulatory Commission regulates our industry.

There is nothing more important to the industry than a credible, effective regulator. It helps us with public perception, it helps us in the political arena, and it helps us operate our plants better. And to their great credit, the NRC commissioners have shown the necessary leadership to reform the regulatory process.

The means to a safer and more effective regulatory system is the new safety-focused, performance-based approach being pilot tested by the commission.

As the industry and the NRC gain experience with the new reactor oversight and assessment process, the approach will be modified and applied to other regulatory areas.

Another reason for the new attitude within the industry is the awareness that the environmental benefits of nuclear energy have potentially enormous economic value. The Clean Air Act sets allowable concentration levels for pollutants such as sulfur dioxide, particulate matter, and nitrogen oxide. That places limitations on new generation and industrial development, particularly in areas that are already out of compliance. Nuclear energy is valuable in offsetting the emissions from fossil fuels and other emission-producing activities and should benefit economically in emissions trading markets.

Emission caps under the Clean Air Act are becoming increasingly restrictive. And so are Environmental Protection Agency rulings. Let me give you a recent example. The EPA recently challenged a state for issuing air emissions to four gas plants. Those plants were to be located near a city whose air quality did not meet standards. The emissions from each separate plant would have been low enough to avoid special review. But the EPA judged that the cumulative effects of those plants had to be considered.

The industry today is also more aggressively seeking recognition of the fact that the total life cycle of nuclear energy makes this form of generation look very good in comparison with any other form of generation. And that includes waste management—which, frankly, has been admirable.

POLITICAL SUPPORT

Now, let me turn to political support.

Key policymakers are becoming increasingly aware of nuclear energy and its environmental benefits.

For example, in September, the Congress passed—and the President signed—the Energy and Water Development Appropriations bill that reflects that support. Included in the fiscal year 2000 budget for the Energy Department is 22.5 million dollars for the Nuclear Energy Research Initiative and 5 million dollars for the Nuclear Energy Plant Optimization program. The latter program is a new program at the Energy Department. And in this climate where Congress is eliminating program budgets—and new program starts are as popular as campaign finance reform bills—this action reflects increased legislator awareness and support for nuclear energy is a vital component of the U.S. energy portfolio.

There also is tremendous interest among members of Congress and key staff members in formalizing caucuses dedicated to examining nuclear technology issues. In the past year, NEI has coordinated seven briefings on nuclear technology issues for staff from the House of Representatives, and attendance has increased exponentially from about 15 at the first meetings to 45 at the most recent meeting—a luncheon with Hans Blix.

In all, representatives from 70 House of Representatives offices have attended these sessions. And, Rep. Joe Knollenberg of Michigan is working on establishing a formal nuclear issues caucus for members of the House. There are similar activities planned for the U.S. Senate as well.

More and more we are hearing favorable policymaker statements, such as the following from Florida Senator Bob Graham: "As we enter the 21st century, it is imperative that our national energy supplies come form a variety of sources...Over the past quarter century, nuclear energy has done more to prevent air pollution than any other form of electricity generation."

At the state level, there is growing awareness of the essential role of nuclear power in meeting Clean Air Act requirements, and the continued importance of emission free nuclear energy in the future. Just ask the state of Georgia, which is concerned about the loss of hundreds of millions of dollars in federal highway funding if it cannot meet federal limits on nitrogen oxides during the summer ozone season.

In the international arena, it is instructive to see how much has changed since the first United Nations meetings on global climate change. Nuclear energy proponents were first viewed as interlopers at the discussions. Today, the presence of proponents and the need to consider nuclear energy are acknowledged. In fact, the International Nuclear Forum was provided an opportunity to speak about the importance of nuclear power to the delegates to the United Nations summit earlier this week in Bonn.

The fact is that, without nuclear energy, any hope of achieving the carbon dioxide levels envisioned by the Kyoto Protocol—or significantly improving the air quality in the U.S. or Europe—are severely diminished. In the United States, we would have to double the reductions of carbon without nuclear power to meet the Kyoto goals by the 2007-2012 timeframe.

PUBLIC SUPPORT

So we have an increasingly positive attitude inside the industry and greater policymaker support. But what about the public?

I believe I can safely say that most of us support the continued use of nuclear energy and wish that the majority of Americans shared our view.

The reality is that Americans who favor nuclear energy clearly outnumber those who are opposed. Let me give you results of national public opinion polls that were conducted for NEI.

NEI regularly conducts two different opinion tracking polls:

- For several years, we have conducted national surveys among representative samples of college graduates who are registered to vote. We chose to examine this population segment because it is politically influential. The sample size in the most recent survey, in March 1999, was 500.
- We also conduct comparison surveys with a nationally representative sample of 1,000 members of the total adult public. The most recent general public survey was October 22-25—that is, about 3 weeks after news coverage of an accident at a Japanese fuel plant. In the interest of time, I will give you only the most recent numbers, but I can tell you that the data before and after the accident are practically identical.

Let's first look at opinions of the two groups—college graduates who are registered to vote and the general public—on key policies for nuclear energy:

- 79% of the general public and 87% of college-educated voters agree that we should renew the licenses of nuclear energy plants that continue to meet federal safety standards.
- 60% of the general public and 73% of college graduates who are registered to vote agree we should keep the option to build more nuclear power plants in the future.
- Finally, 42% of the general public and 52% of college-educated voters agree we should *definitely* build more nuclear energy plants in the future. Even those numbers are respectable.

Now let's look at feelings about nuclear energy. Both the general public and college graduates who are registered to vote favor the use of nuclear energy by a two-to-one margin. See Table 1.

Table 1. Do you *strongly favor, somewhat favor, somewhat oppose,* or *strongly oppose* the use of nuclear energy as one of the ways to provide electricity for the United States?

	College Grads/Registered to Vote	Public
Strongly favor	23	25
Somewhat favor	39	37
Somewhat oppose	23	16
Strongly oppose	14	14
Don't know	1	8
Favor	**62**	**62**
Oppose	**37**	**30**

These numbers are already high, even though awareness of nuclear energy's environmental benefits is just beginning to take hold. Listen to what happens to attitudes when people are reminded of these environmental advantages. I'll tell you about the numbers for the college graduate/registered voter group—but they are the same in all the surveys where we have tested the impact of the environmental message.

After asking the question about favoring or opposing nuclear energy, we then provided one sentence of information and asked the question again. The sentence of information said:

"There are more than one hundred nuclear energy plants in the United States that generate one-fifth of all the electricity we use in the United States without emitting any greenhouse gases or other air pollutants."

Just this one sentence of information increased the number favoring the use of nuclear energy by 12 percentage points—from 62% to 74%.

Table 2. Before and after reminder about environmental benefits: Do you *strongly favor, somewhat favor, somewhat oppose,* or *strongly oppose* the use of nuclear energy as one of the ways to provide electricity for the United States? (College Graduates/Registered to Vote)

	Before	After
Strongly favor	**23**	**34**
Somewhat favor	39	40
Somewhat oppose	23	16
Strongly oppose	14	9
Don't know	1	1
Favor	**62**	**74**
Oppose	**37**	**25**

Also, the number strongly in favor of nuclear energy increased sharply and the number strongly opposed dropped by 30 percent.

Public awareness of nuclear energy's environmental benefits is increasing. An open-ended question gave respondents an opportunity to name up to four advantages of nuclear energy. In March 1999, 40% mentioned an environmental advantage, compared with 30% in January 1998.

THE PERCEPTION GAP CHALLENGE

Public opinion is quite favorable and promises to become more so as public and policymaker concern for the environment increases. Unfortunately, a lot of people do not know that.

In survey after survey, we find that a majority of respondents personally favor nuclear energy but only about 20% or so believe that a majority of the public in their community shares their view. This misperception about public opinion is at least partly due to the fact that opponents of nuclear energy have typically been more outspoken than proponents. It's a problem, especially when policymakers hold this misperception, because it inhibits their willingness to take actions that they believe to be right but unpopular.

Over time, with the industry's new attitude and the growing support among policymakers, this misperception about public opinion can be reversed. We need to keep communicating the benefits of nuclear energy and demonstrating the support that exists in order to keep up the positive momentum. In addition to traditional industry voices, I am happy to report that NEI is committed to supporting strong advocates within its en*Act* grassroots network and the Alliance for Sound Nuclear Policy, the development of Nuclear Young Generation-North America and U.S. Women in Nuclear.

Let me close by sharing with you a quote some years ago by columnist Ben Wattenberg that was in an article celebrating the public's rediscovery of a number of seemingly forgotten, but manifestly good ideas. Ben—known as the Connecticut Avenue philosopher—said "There is nothing so powerful as an old idea whose time has come again."

Nuclear energy is an old idea whose time as come again. As we stand on the cusp of the new millennium, we must recapture the power of that idea to ensure that nuclear energy—and the many benefits it provides society—never fades into the background of the American consciousness.

NUCLEAR POWER: LIABILITY - OR ASSET?

Myron B. Kratzer

Consultant
25419 Galashields Circle
Bonita Springs, FL 34134

INTRODUCTION

If a beneficent but supremely mischievous deity decided to provide the human species with a new, environmentally benign and virtually inexhaustible source of energy, just as the classical sources were approaching serious depletion and as billions of additional people were demanding their fair share of what remained, there is no doubt whatsoever what he would come up with. He would devise an energy source whose use is accompanied by radiations that are regarded as mysterious and are widely feared, even though present since the dawn of life or earth; he would see to it that there is a small potential for accidents that could disperse this radiation, even though such accidents would be avoidable and containable by good engineering. He would ensure that the energy source is not completely consumed through routine use, and indeed can even be augmented, leaving a valuable energy resource in its place, but one which can easily be mistaken for trash. Finally, in an unparalleled burst of creative mischief, he would endow the energy source with features that allow it to be fashioned into the most destructive weapon ever devised, and would arrange for its discovery and that of the basic process for its use, not in peaceful research but in the course of the development of that very weapon.

In what way and how much do the military origins and uses of fission energy impact the prospects for revival of the nuclear power option? Are they a serious impediment; are they of little significance; or is it just possible that, if fully understood, the military implications are a positive factor? No assessment of the future of nuclear power can be complete without consideration of the military use issue, the essence of which is the potential spread of nuclear weapons to additional countries or even subnational entities. This paper reviews this issue, giving particular attention to international nuclear safeguards, certainly the most distinctive, and probably the most misunderstood feature of the nuclear nonproliferation regime.

The Challenges to Nuclear Power in the Twenty-First Century
Edited by Kursunoglu *et al.*, Kluwer Academic/Plenum Publishers, New York, 2000

HISTORICAL BACKGROUND

Our concern today is with the future, but, as always, the past can be instructive, so let's take a brief look at the history. The first thing we see is that the proposition that the military potential of nuclear energy might be a positive factor favoring its peaceful development is not so farfetched after all. Nothing is more fundamental to an understanding of the proliferation issue than the simple fact that, just as fission energy was discovered in response to military incentives, so can it survive regardless of the future of peaceful uses. The "Atoms for Peace" program, while still maligned by anti-nuclear activists, was designed not to spread peaceful nuclear technology, as its detractors sometimes contend, but to take advantage of the appeal of peaceful uses to gain a measure of control and restraint over what would otherwise have been an explosion of uncontrolled national nuclear programs.

"Atoms for Peace" was once referred to as the "mindless Atoms for Peace program" by officials of the Carter Administration, but this characterization conveniently overlooked the fact that proliferation by that time - and still - had been far more limited than predicted, and that the program enabled the creation of an unprecedented international structure designed to avoid proliferation - the nonproliferation regime. Today, there is no serious dispute that this regime, including the nonproliferation treaty which is its cornerstone, could never have been brought into being without the quid pro quo of peaceful nuclear technology shared through Atoms for Peace.

It is an interesting historical footnote, and surprising in hindsight, that at its inception Atoms for Peace required no commitment on the part of recipient nations to forego nuclear weapons development. The prohibition against military use applied only to the specific assistance provided under the program, the thinking being that this assistance would prove sufficiently attractive to persuade recipients that it should not be jeopardized by engaging in parallel nuclear weapons programs. By and large, this assumption proved correct, and there is no evidence that the nations that still remain outside the regime or, worse yet, have deliberately violated it would have followed a different course if a prohibition against military use had been on of the original conditions of Atoms for Peace. Nevertheless, the explicit quid pro quo of the NPT provides a far more comfortable basis for proceeding with nuclear power cooperation.

A central feature of the Carter policy was that the nuclear power fuel cycle as it was then generally conceived; that is, with reprocessing and recovery and recycle of plutonium, constituted the greatest threat of proliferation. Once again, this ignored history, since no nation had relied on material from its civil nuclear program for the development and manufacture of its initial nuclear weapons. This remains true today, although material for expanding nuclear weapons programs has certainly been derived from dual-purpose facilities generating nuclear power, just as material for peaceful use, notably enriched uranium, has been derived from facilities initially built to meet perceived military needs.

Although the Carter policy, largely for nonproliferation reasons, was inherently inimical to nuclear power, which it characterized as "a last resort," it did not directly reject nuclear power per se. To have done so would have been too blatantly inconsistent for an administration which believed, not entirely without reason, that it was facing an energy crisis that was "the moral equivalent of war." Instead, the Carter policy reserved its antipathy largely for plutonium recycle, and especially for reprocessing, plutonium, and the fast breeder. By doing so, it initiated a dispute on fuel cycle policy that has had continuing repercussions and will no doubt continue to influence fuel cycle policy for the foreseeable future.

What can we distill from this background of relevance to the future?

First, the development of nuclear weapons preceded peaceful nuclear development; it does not depend on it in any way; and it can continue and spread to additional countries even if nuclear power is abandoned globally or is absent in proliferant countries.

Second, typically small, dedicated facilities and not nuclear power plants or their associated fuel cycle facilities have been the source of nuclear material for all proliferation to date and are the most probable route to any future proliferation.

Third, excesses beget excesses. While many nuclear advocates responded to the exaggerated importance placed on the fuel cycle by the Carter policy by rejecting any relationship of nuclear power to proliferation, this view is equally unsupportable. Although less likely, the fuel cycle - with or without recycle - can be a source of nuclear material for weapons. This is why safeguards and many other features of the nonproliferation regime were developed.

THE CURRENT SITUATION

Where does nonproliferation stand in the myriad of institutional, regulatory, public acceptance, economic and, at times, even technical problems now affecting the future of the nuclear option? Is the potential contribution of nuclear power activities to further proliferation seen by policy makers and, of equal importance, the public as a significant obstacle to its revival? Or is it viewed as a potential benefit capable of reducing proliferation risks.

Although there has been relatively little in-depth assessment of public attitudes on the proliferation, the information available is revealing. When asked to name any disadvantages they perceive of nuclear energy as a source of electric power, only 1% of a sample of U.S. college graduates named its weapons or proliferation potential, among more than ten factors, of which accidents and danger rated highest, at 40%. Another relevant poll which focussed explicitly on the issue of the disposition of surplus weapons plutonium, showed strong support for burning this material as reactor fuel even when compared directly with the alternative of immobilization in glass, with the MOX option favored by 80%. These data, for which we are indebted as always to Dr. Ann Bisconti of Bisconti Research Inc., suggest that the public is ahead of many policy makers in this area, and that, at least in the U.S. where surplus weapons plutonium is available, the weapons aspects are viewed by the public as a reason to favor nuclear power rather than a reason to oppose it. Although public opinion data on this issue from other countries may not be available, the U.S. data are conclusive enough to suggest that there is a good chance that there would be support elsewhere for burning excess weapons plutonium in national reactors. Indeed, the historic logic of contributing directly to the destruction of nuclear weapons material might well be compelling in other countries, especially Japan.

Although some progress is being made in pursuing the MOX disposition option in the United States, the obstacles to its eventual implementation, especially in the licensing area, are daunting. Fabrication and irradiation of at least initial cores in Europe, as strongly recommended by Senator Domenici, could by-pass many of these obstacles, while accelerating the start of the program by two or more years. European objections that undertaking these activities could result in a corresponding delay in working off existing stocks of European reactor-grade plutonium and would thus be of questionable net benefit are understandable. Much if not all of this displacement, however, might be avoided through careful scheduling, and even some displacement would be acceptable if it allows the initiation of a disposition program that would otherwise be indefinitely delayed.

Disposition of the surplus weapons plutonium in the U.S. and Russia remains a critically important international security objective, and the MOX option is the only one capable of attracting Russian support. Reducing the stocks of this material is important not only in its own right, but as a concrete demonstration of the ability of responsible governments to make and implement policy which allows nuclear power to contribute, as originally foreseen, to the reduction of proliferation risks. The responsible U.S. government agencies have so far been hostile to international fabrication and irradiation, perhaps out of concern that undertaking these activities abroad would result in dilution of program control. But this need not be an either/or situation and there is no substitute for getting started. At a minimum, the U.S. government agencies involved owe top policy makers and the Congress an explanation for their opposition to direct international participation in the MOX disposition program.

The support of the current U.S. Administration and, indeed, of Western governments generally for nuclear power remains limited to say the least, as the curious silence at Kyoto so vividly illustrates. Officially, the early policy statements of the Clinton Administration indicate that there is no objection to the once-through fuel cycle on nonproliferation policy grounds. U.S. support for LWRs in North Korea on the basis of strict adherence to a once-through fuel cycle reaffirms this position. Notwithstanding this official stance, it seems likely that the military heritage and linkages of nuclear power play some role in keeping official support for the nuclear option at the "last resort" level.

If nonproliferation considerations have not led to official opposition to nuclear power, their effect on fuel cycle policy has been profound. Although, its rhetoric and many of its implementating actions have been more restrained, the Clinton Administration has, in principle, adopted the Carter policy of opposition to reprocessing and plutonium recycle. In at least one important area, however, it has inexplicably out-Cartered earlier policy by terminating work on proliferation-resistant fuel cycles that involve recycle of still highly radioactive plutonium.

The issue of plutonium recovery and recycle cannot be separated from that of spent fuel and its management or disposal. In this respect, in a curious role reversal, the traditional anti-nuclear activists have chosen to understate the proliferation aspects of spent fuel, the waste form they advocate, largely, it can be assumed, in order to avoid acknowledging that spent fuel is a valuable energy resource and that reprocessing is a much lower proliferation barrier than they have represented. The result has been the adoption of the U. S. policy that spent fuel is nuclear waste that must be permanently disposed of in geologic repositories. The acquiescence of the U.S. nuclear industry in the characterization of spent fuel as waste, while understandable in terms of seeking a resolution of the spent fuel issue, has been an indispensable element in the adoption and maintenance of this policy and the consequent impasse that threatens the continued viability of the U.S. nuclear enterprise.

While public understanding of nuclear issues may lack sophistication and is often based on inadequate or even misleading information, the public's assessments are not irrational. Having been told over many years that spent fuel is nuclear waste, it is only natural that the public should insist on its disposal. If and when effectively informed of the fact that spent fuel is not a waste but an energy resource, there is every reason to believe that the public will reject its deliberate burial and favor its storage under secure conditions, just as it now favors consuming, rather than immobilizing, surplus weapons plutonium.

The spent fuel issue is central to long-term fuel cycle policy, not simply because large volumes are threatening to clog the arteries of the nuclear power industry but because spent fuel is the repository of most of the world's plutonium, some 1000 tons at present, and is already dispersed among the 30-odd countries in which nuclear power plants are located. The indefinite accumulation of these dispersed inventories has proliferation implications that are at least comparable in their gravity to the surplus weapons plutonium inventories in Russia.

The report of the American Nuclear Society's International Panel on Protection and Management of Plutonium - the Seaborg Panel - was the first to emphasize the importance of degrading the isotopic composition of surplus weapons plutonium through irradiation and the Panel had the question of spent fuel right as well. The long-term goal of fuel cycle policy, the Panel concluded, must be to bring the consumption and production of plutonium into balance, and to reduce the inventories of plutonium in all its forms to the minimums consistent with efficient operation. This goal does not, as the report emphasized, call for immediate resumption of reprocessing in countries, such as the United States, where it has been delayed, nor does it call for or countenance reprocessing in every nation generating power. What is called for, however, is a recognition that since spent fuel is a valuable resource, its retrievable storage until decisions are made on its disposition is not a stop-gap, but a rational and, indeed, preferable fuel cycle approach in its own right.

Another implication of adopting the goal of balancing plutonium production and consumption and minimizing inventories is that every effort should be made to reduce the number of countries in which plutonium is stored by finding sites in appropriate locations that will accept spent fuel from other nations. In seeking to locate such sites, countries that are not necessarily engaged in nuclear power generation themselves should not be excluded, nor should all countries considered to fall into the category of "developing." Nations that decide to pursue nuclear power are normally regarded as welcome to do so provided they have the capability of making, and implementing this decision on a mature and well-informed basis, and are normally given assistance by the IAEA and others in making the necessary assessments and preparations.

Nations that decide that they wish to limit their participation in the nuclear field to spent fuel storage, or, for that matter, to combine spent fuel storage for others with nuclear power generation, are entitled to no less consideration. The a priori judgment of the environmental and developmental elitists that all "developing" countries are incapable of making such decisions for themselves and should be excluded from this activity is at least as patronizing as is improperly inducing unprepared countries to provide such a service. The fact that storage would be interim in nature further supports the propriety of allowing countries capable of making independent and mature judgments to provide this service if they wish to do so.

While spent fuel inventories, and with them their plutonium content, are continuing to grow, significant amounts of spent fuel are being reprocessed both from reactors located in the few nations with suitable reprocessing facilities and from reactors located in other countries. If the goal of balanced production and consumption of plutonium is adopted and pursued, much greater quantities of spent fuel will have to be reprocessed in the future, but the goal should be to confine this activity to as few nations of unquestioned nonproliferation credentials as possible. Ideally, fabrication of the corresponding MOX or other recycle fuel should be limited to the same countries in which reprocessing takes place. A key question will be whether irradiation of the plutonium fuel will be permitted elsewhere and, if so, under what conditions. Finally, reprocessing should employ proliferation-resistant technology in which plutonium is never completely separated from its protective barrier of radioactive fission products. Meeting this goal requires that research and development on proliferation-resistant fuel cycles be encouraged and that international cooperation in this area be permitted. It will also require the development of reactor technologies capable of repeated recycle leading to complete consumption of actinides.

Attainment of the goal of balanced plutonium production and consumption and minimum inventory will require a number of decades. Paradoxically, achievement of the goal will be most important in the event that nuclear power is phased out, leaving large and increasingly accessible plutonium inventories in many countries if no provision for their destruction has been made. While the global abandonment of nuclear power is extremely unlikely, termination in one or more countries is not improbable. If this occurs, transfer or reprocessing of any spent fuel inventory in the countries concerned would be desirable.

INTERNATIONAL SAFEGUARDS

Safeguards, the term employed in the IAEA Statute for what is customarily called verification in other treaties, are the most distinctive feature of the nonproliferation regime. They represent the first, most widely applied, and, in some respects the most successful implementation of the post World War II principle that compliance with important international security obligations should not be assumed and should be verified through independent, objective means. Following the initiation of international nuclear cooperation in the late 1950s, the peaceful use undertakings of bilateral Agreements for Cooperation were generally verified by the nations, most frequently the U.S., supplying nuclear materials or equipment to other countries. By the mid 1960s, however, the responsibility for the application of safeguards under bilateral Agreements for Cooperation had been successfully transferred to the IAEA, except in the European Community States, where safeguards were applied by Euratom.

The conclusion of the NPT in 1968, and its coming into force in 1971, assigned a major new responsibility, as well as a greatly increased workload, to the IAEA. This was the application of safeguards not only to specified activities and materials supplied bilaterally, but to treaty States' entire peaceful nuclear programs (since the NPT prohibited non-nuclear weapons States from undertaking nuclear weapons programs, this was tantamount to the application of safeguards to all of the States nuclear activities, except for non-proscribed non-weapons military programs, a circumstance which has never arisen). In the unambiguous words of the IAEA's NPT safeguards system, safeguards agreements are to " provide for the Agency's right <u>and obligation</u> to ensure that safeguards will be applied...on <u>all</u> source or special fissionable material in <u>all</u> peaceful nuclear activities within the territory of the State." (emphasis added.)

The discovery following the Gulf War that Iraq, an NPT party, had been engaged in several major programs devoted to the development of nuclear weapons, which had escaped detection or even investigation by the IAEA shattered the complacency that had developed in respect to IAEA safeguards, and focussed attention on the fact that the Agency had confined its safeguards activities to countries' "declared" nuclear activities, notwithstanding the language, cited above.

The reasons for the Agency's self-imposed limitation of its NPT safeguards to declared activities in the face of the unambiguous requirements of the safeguards system developed and adopted by its member States are complex and troubling, but are outside the scope of this paper. The fact that this occurred, however, raises legitimate questions as to the nature and effectiveness of the oversight of the Agency's safeguards program on the part of both the Agency's own organs and its member states individually. It would be appropriate to address these questions during the course of the current intensive effort to strengthen the effectiveness of the Agency's safeguards system. This consideration must be based on the principle, often overlooked, that Agency member states concerned that there be full compliance with safeguards agreements have just as much right to insist on effective safeguards as inspected States have to insist on efficient implementation.

The response of the Agency and its Board of Governors to the Iraqi set-back was, in fact, both prompt and effective. In little more than six months after conclusion of the Gulf War, the Board had reaffirmed that the Agency's right and obligation to apply safeguards to all activities extended to concealed or "undeclared" activities as well as those that were declared; that the "special inspections" that the Agency was authorized to conduct under NPT safeguards agreements allowed it to have access to any location in a State in connection with the application of safeguards to undeclared activities; and that the Agency had the right to make use of all information available to it in applying its safeguards and reaching safeguards conclusions. The effectiveness of these decisions was strikingly demonstrated in the Agency's application of safeguards to North Korea's nuclear program. Soon after initiation of these safeguards, the Agency discovered discrepancies which established that the initial inventory of nuclear material declared by North Korea was incomplete and, acting on information provided by a member State, that North Korea had concealed locations at which undeclared nuclear material was likely to be present. Following unproductive consultations with the North Korean government, the Agency demanded special inspection access to these locations and, on denial of this access, reached and reported to the UN Security Council, a finding of non-compliance on the part of North Korea.

The Agency's finding of noncompliance set in motion a remedial process leading to the so-called "Agreed Solution" under which North Korea will receive two light-water reactors while the Agency monitors a freeze on its former weapons-oriented nuclear program. While opinions may differ as to the appropriateness of this solution, the remedial process following findings of non-compliance outside the Agency's statutory competence and the solution adopted in this case in no way detracts from the fact that North Korea was a safeguards success story. The success of Agency safeguards in this endeavor remains of enormous importance by demonstrating that the Agency's existing NPT safeguards agreements provide it with the basic rights and mechanisms needed to deal with undeclared nuclear activities undertaken by NPT States. While further improvements in these rights and mechanisms, discussed below, have been made, it is essential that the availability of these improvements not be allowed to make the Agency's strong existing rights fall once again into disuse.

Following its key reaffirmation of the Agency's right to seek and apply safeguards to undeclared activities, the Agency's Board also decided that further strengthening measures would be desirable, setting in motion the program known as 93+2. The development and implementation of safeguards, particularly during the long period of complacency that ended with the post-Gulf War discoveries in Iraq, has by and large been the province of safeguards specialists in the Agency and member governments. However, the plans for implementation of the new rights and particularly the manner in which these new rights interact with and are integrated with the Agency's existing safeguards rights and system are now being developed. This process, which has come to be known as integration, or integrated safeguards, has profound implications for the ability of the IAEA to meet its statutory safeguards obligations. Accordingly, increased attention by policy makers not only within the Agency but in member governments is now essential.

A word of explanation is in order here. The Agency's new rights are intended to improve the Agency's assurance that no undeclared nuclear activities are taking place in an inspected nation - a key failure in the case of pre-Gulf War Iraq. The new rights designed to improve this assurance do not provide the Agency with access to additional locations, since the Agency's special inspection rights already provide it, under specified circumstances, with access to any location within a State. The new rights do, however, lower the threshold which must be crossed before the Agency can assert their access rights. They also increase the Agency's opportunities to make use of a valuable new safeguards technique, environmental sampling, and they increase the amount of information on its nuclear program that each nation is obliged to routinely provide the Agency. All of these new rights are desirable. There is no disagreement, in principle, however, that the implementation of these rights will not and cannot provide the Agency with complete assurance of the absence of undeclared activities, and that the level of this assurance will always be less than the assurance that the Agency is capable of securing and normally does secure that declared nuclear material is not being diverted. It is inherently difficult to prove a negative - in this case, that undeclared nuclear activities are not taking place.

The Agency's new rights will not apply automatically to each country in which the Agency is already applying its NPT safeguards system; they will apply only to those countries that have voluntarily entered into an addendum, or protocol, to its existing NPT safeguards agreement with the Agency. A number of the countries that have taken this step or indicated their intention to do so have taken the position that as the Agency implements its rights under the protocol, thereby improving its assurance of the absence of undeclared activities, it can and should reduce the intensity of its classical safeguards activities designed to verify the non-diversion of declared nuclear material. The Agency's own safeguards staff generally agrees with this concept, although many questions and differences remain as to the degree to which classical safeguards measures can or should be reduced. While some governments or their experts base their support for this "trade-off" in safeguards to what they see as a logical relationship between the assurance of absence of undeclared activities and the assurance of non-diversion of declared nuclear materials, other governments or experts appear to believe that giving the Agency new rights to seek undeclared activities should be rewarded by a decrease in classical safeguards inspections, regardless of any "logical" connection. Although this "quid pro quo" view is generally rejected in principle both within the Agency and among member states, there is a strong presumption of "trade offs" that, in practice and in its results, may differ little from the "quid pro quo" concept.

Notwithstanding the Iraqi "lesson learned" that the possibility of undeclared nuclear activities must be taken seriously and their possible existence sought out, the concern with undeclared activities as a proliferation risk is not new and their possible existence has always been recognized, indeed, presumed, in any serious analysis of safeguards. Even purified plutonium or highly enriched uranium metals are "harmless" in bulk form. Further steps, specifically fabrication into weapons components, are necessary before these materials can result in proliferation and these steps, while perhaps not demanding, are not trivial. They are necessarily presumed to exist if the diversion of separated plutonium or HEU is discovered, since no reliable means for their detection are available.

The difference between the diversion of plutonium or HEU, referred to in safeguards terminology as direct-use materials, and the diversion of other nuclear materials is one of degree. Upgrading is always necessary before diverted material becomes weapons components, and the magnitude of this difference, perhaps even its sign, is by no means always obvious. Although the Carter Administration placed great emphasis on reprocessing as a proliferation barrier, and attempted to dismiss the possibility of "quick and dirty" reprocessing on technical grounds, reprocessing experts generally agree that reprocessing can be done on a small scale cheaply and easily if safety and efficiency are downgraded. From the technical point of view, perhaps the most important change in the safeguards environment in recent years is the availability of enrichment technology, which allows a wide spectrum of countries to successfully undertake small-scale enrichment, as the experience of countries such as Pakistan, Argentina, and Brazil demonstrates, and Iraq nearly confirmed.

On a small scale and with some efforts at concealment, the detection of reprocessing and, even more so, of enrichment by means available to the IAEA or even by national technical means, is difficult and by no means assured. The safeguards lessons to be learned from Iraq did not end with the post-Gulf War discoveries of undeclared activities. On the contrary, the extreme difficulty of developing assurances of the absence of undeclared activities and materials, even in a country of modest size, modest technological capabilities, and subject to an inspection regime of unprecedented rigor, when that country, like Iraq, is bent on concealment is a "lesson to be learned" on a continuing basis for more than seven years in Iraq.

The relevance of this background to the current "integrated safeguards" debate is clear. If "upgrading" activities, whether weapons component fabrication, enrichment, reprocessing, or any other, can be ruled out with certainty, then classical safeguards designed to detect the diversion of declared nuclear materials could, in principle, be eliminated. However, both logic and experience demonstrate that these activities cannot be ruled out with anything approaching certainty, making continued application of effective safeguards for the detection of diversion imperative. Indeed, one of the most useful measures for improving the assurance of the absence of undeclared upgrading activities is verifying the non-diversion of declared nuclear materials, since such diversion implies the existence of undeclared activities.

It is often said that the lesson of Iraq was the importance of improving the detection of undeclared nuclear activities, but this misreads the Iraqi experience and what was to be learned from it. One of the indispensable requirements of effective safeguards is completeness. In addition to its undeclared activities, Iraq was also engaged in diversion of declared material, and safeguards confined to the detection of undeclared activities but overlooking diversion of declared materials would have been every bit as defective as the reverse situation that prevailed. Safeguards that effectively cover some proliferation pathways, while leaving others uncovered may make proliferation by the covered pathways difficult, while providing a free ticket for the uncovered pathways or strategies. Only those safeguards are effective that introduce a significant risk of detection in all credible diversion pathways or proliferation strategies.

The second requirement of effective safeguards is independent verification. Here, too, a word of explanation is necessary. Two basic approaches to the investigation of wrongdoing are possible. In the case of the everyday activities of individuals and many organizations, at least in the countries that most of us live in, compliance with the law is presumed and investigation of non-compliance takes place only when triggered by some evidence of wrongdoing by the person or persons concerned. However, when persons engage in inherently risky activities - driving a car, practicing medicine, building and operating a nuclear power plant just to name a few examples - compliance or competence must be affirmatively demonstrated through licensing and enforcement activities. In domestic terms, this is referred to as regulation. In arms control terms, it is referred to as verification.

Since the earliest studies of proliferation, as well as many other nuclear arms control issues, a broad consensus has existed that compliance should not be presumed and must be verified. In the NPT, as well as relevant bilateral and multilateral agreements, verification is explicitly required, and this requirement is reflected in the Agency's NPT safeguards system. Moreover, this verification must be independent, that is, it must be based on information acquired or verified through the Agency's own measures. Information provided by inspected states, while helpful and in some circumstances even essential, cannot be presumed to be complete or accurate unless and until verified. In fact, verification can mean only ascertaining the truth through independent means, and the term "independent verification" is a redundancy that has come into use to provide emphasis and avoid misunderstanding.

While no final decisions have been made, certain of the approaches currently being considered for "integrated safeguards" have the potential for falling short of either or both of the indispensable requirements of meaningful safeguards: completeness and independent verification. For example, substantial reductions in safeguards on indirect-use material are being advocated, on the grounds that increased assurance of the absence of undeclared activities make safeguards to detect diversion of indirect-use material of much less importance.

As another example, increased reliance on the control activities of national systems is being considered, with the potential for serious departure from the requirement of independent verification. In their place, it is suggested that the Agency's new rights will allow it to accomplish the same purpose as classical safeguards, through increased investigation when "inconsistencies or questions" arise. The similarity of this approach to the investigative system of law enforcement and its departure from verification are evident.

Fortunately, the world is not full of Iraqs or North Koreas. Countries that are not involved in overt security disputes with their neighbors or others are, in all probability, not engaged in clandestine nuclear weapons programs. Conversely, countries developing nuclear weapons, clandestinely or otherwise, typically have signaled these activities through policy statements or other means. Three threshold countries whose names are familiar to all were long widely believed to possess nuclear weapons, but at least had the good manners to refrain from bad faith signature of the NPT. Two of these have recently confirmed their weapons programs through testing.

In short, a case can be made for relying on safeguards more akin to investigation than to verification, depending on a strong, and usually correct, presumption of compliance in the absence of indications to the contrary. Some national safeguards officials specifically advocate taking into account such political factors as "societal openness" in making judgments in the still-to-be-developed integrated safeguards. Others, citing the general understanding that international organizations are foreclosed from giving weight to political considerations, would avoid formal reliance on such factors, but seem nevertheless to be prepared to take them into account. There is no doubt that clandestine nuclear programs are less likely to take place in "open societies," and more likely to be revealed if attempted. At the same time, it cannot be overlooked that governments of even the most "open societies" maintain elaborate programs to classify and protect information they deem to be of national security importance, and these programs are sometimes successful, even in the United States.

Information from all sources, including inspected countries themselves, is of value to safeguards. What must be certain, however, is that information provided by an inspected country will never knowingly include information on activities that the country wishes to conceal, and this self-evident fact applies with greatest force to the information specifically called for by and prepared for submission to the IAEA safeguards system. Mistakes are, of course, possible and information pointing to the existence of undeclared activities may inadvertently find its way into public documents. The verification system of the Chemical Weapons Convention CWC) places considerable emphasis on reporting by treaty parties, but many Chemical Weapons precursors have non-weapons uses and are items of commerce, making deliberate omission more difficult and inadvertent mistakes more likely. Nuclear materials, in general, have no significant non-nuclear uses, and inadvertent disclosure is likely to less likely occur than in the case of the CWC.

The basic point is that, as a tool for the detection or even raising suspicion of undeclared activities, the provision of information by inspected states on their own activities, no matter how carefully and detailed requests are framed, must be of limited value.

Central to the issue of whether or to what degree a shift toward investigatory and away from verification safeguards would be justified and acceptable are the costs and other burdens of the existing "classical" safeguards system. Departing from this system, which has by and large demonstrated its effectiveness in verifying declared peaceful nuclear activities, might be acceptable if its costs and intrusiveness were major burdens. Departing from the existing system if this condition is not met would raise very serious questions.

It is a truism, of course, that inspection and inspectors are not popular on the part of those being inspected. Nevertheless, by objective standards, it is difficult to avoid the conclusion that the financial costs and other burdens of safeguards, for complying nations, are at most quite modest. If the nuclear power produced by the some 200 reactors under safeguards is valued at 4 cents per kilowatt hour, its total value would be of the order of $50 billion annually. Agency safeguards costs of some $100 million annually would represent about 0.2% of this value, virtually within the "noise level." Intrusiveness is a subjective criterion, for which no agreed definition exists, but much of what is termed intrusiveness are measures that inspected states take to avoid or minimize safeguards. For countries in compliance with their nonproliferation undertakings, many of these avoidance measures seem unnecessary. Compared to the presence of scores of domestic security personnel typical of nuclear installations the occasional visits of Agency inspectors, or even the minimal continuous presence at facilities such as reprocessing plants seems unexceptional.

Despite this background, efforts to limit, and even scale back, IAEA safeguards have been endemic, and these efforts are currently undergoing one of their peaks, taking advantage of the Agency's new safeguards rights under the Protocol and the process of devising "integrated safeguards."

The benefit side of the safeguards equation must also be considered. Putting aside the incalculable value of detecting non-compliance that might otherwise escape detection, the verification of compliance on the part of the vast majority of States who would be in compliance with their nonproliferation undertakings even if verification safeguards did not exist is an essential element in setting the climate that permits peaceful nuclear activities, and especially international cooperation, to take place. No one benefits more from safeguards than the countries engaged in peaceful nuclear activities, especially nuclear power generation, but among these countries are several that have traditionally exerted strenuous efforts to limit the scale and quality of the Agency's safeguards system.

Recent developments to strengthen the Agency safeguards system in relation to undeclared nuclear activities, notably the Protocol, are valuable and welcome. There are, however, serious downside risks to these otherwise welcome developments. First, unless carefully structured, the integration of the new measures with classical safeguards on declared nuclear activities could lead to an undesirable weakening of the existing system in a manner that would undermine fulfilling the indispensable requirement of completeness. Second, and of equal if not greater importance, concentration on the Protocol could interfere with application of the Agency's existing strong rights to deal with undeclared nuclear activities, which remain of crucial importance, especially in States that do not accept the Protocol.

PHYSICAL SECURITY

Safeguards are the system designed to verify compliance or detect noncompliance with nonproliferation undertakings on the part of the States giving these undertakings themselves. In contrast, physical security refers to the measures employed to minimize the chance that nuclear material would be seized or stolen by subnational actors. Thus, States themselves have the greatest incentive to employ effective physical security systems, and it is well-settled that the conduct of these systems is a State responsibility. Nevertheless, it is apparent that the consequences of a failure of physical security that placed nuclear material in the hands of terrorists or other unauthorized parties, could well have international consequences. Thus, the international community has a strong, legitimate interest in effective application of physical security systems by individual States. This international interest is not adequately reflected in current arrangements.

This issue has been addressed by a number of groups, including the ANS Seaborg Panel. Their conclusion has been that States should be called on to meet specified minimum standards of physical security, and that their performance in doing so should be monitored by an international authority, preferably the IAEA. Unfortunately, despite the general support for the concept, effective action to implement these recommendations has not taken place.

CONCLUSIONS AND RECOMMENDATIONS

1. Proliferation concerns do not play a significant role in limiting the acceptance of nuclear power on the part of the general public. The nonproliferation regime has been effective in limiting proliferation to levels far below those forecast and safeguards, the most distinctive feature of the regime, work. However, there is little doubt that concern with proliferation is a factor in the view of the current and some past U.S. administrations of nuclear power as a "last resort."
2. Proliferation concerns have been and continue to be the basic cause of the official U.S. opposition to reprocessing and plutonium recycle, and have thus led to the official U.S. categorization of spent fuel as "nuclear waste" which should be permanently buried in geologic repositories.
3. In contrast to current U.S. policy, the long-term goal of nuclear policy should be to bring plutonium production and consumption into balance, and to reduce inventories of plutonium in all forms, including spent fuel, to the minimum consistent with the need for working stocks. Provided that plutonium is kept under effective safeguards, this approach is more proliferation-resistant than the current U.S.-favored approach of the once-through fuel cycle and burial of spent fuel in geologic repositories.
4. While the goal of bringing plutonium production and consumption into balance is a long term one, research and development on proliferation-resistant fuel cycles should be taking place at present. International cooperation of the appropriate countries in this R&D is also essential. Failure to pursue a suitable R&D effort and international cooperation is virtually certain to result in the adoption of the most proliferation-prone fuel cycle when the plutonium breeder is deployed in the next century.
5. The number of countries in which spent fuel is stored should be reduced through the development of international retrievable spent fuel storage facilities in one or more countries. Countries that are normally considered "developing nations" and wish to pursue this activity as a means of producing revenue or preparing for more comprehensive nuclear programs in the future should not be foreclosed from doing so, provided that they are capable of mature judgment in the assessment and implementation of such a decision.
6. Implementation of the MOX option for disposition of much of the U.S. and Russian surplus weapons plutonium is an important international security goal. It should go forward with international participation in MOX fabrication and irradiation in order to realize the earliest possible start and avoid potentially prohibitory U.S. political and regulatory obstacles.
7. Effective safeguards, which require completeness and independent verification, are essential to maintenance of a climate of confidence in nonproliferation which allow nuclear power and international nuclear trade to take place.
8. Recent developments dedicated to further improving the IAEA's capability to detect undeclared nuclear activities are desirable and should be implemented. However, undue concentration on implementation of the Protocol could undermine the effectiveness of classical safeguards on declared activities and could further weaken the Agency's ability to make use of its strong existing rights to deal with undeclared activities when necessary, especially in countries that do not accept the Protocol.
9. The integration of Protocol measures with those of the existing safeguards systems should be approached in an evolutionary manner and with great care, in order to avoid weakening the safeguards system.

10. The international community has a legitimate interest in the adequacy of national measures to apply effective physical security to avoid theft or seizure of nuclear material by unauthorized subnational individuals or groups. Assurance of this adequacy through IAEA monitoring under an international convention would be a desirable approach.

REMARKS FOR THE CONCLUDING PANEL

C. Pierre Zaleski

Délégué Général du Centre de Géopolitique de l'Energie et des Matières Premières
Université Paris IX – Dauphine
Place du Maréchal De Lattre de Tassigny - 755775 Paris cedex 16

I would like to address here the issues of non-proliferation of nuclear weapons and of the competitiveness of nuclear power.

NON-PROLIFERATION

The Link between Nuclear Power and Nuclear Weapons

It is acknowledged in many circles, but still not in the non-proliferation «community» and among antinuclear activists, that the link between nuclear power from a well-safeguarded light water nuclear power plant and nuclear weapons is very tenuous. As an example, one can mention the United States proposal to provide North Korea with light water reactors in view to minimize the risk of weapons proliferation by that country, by getting them to agree to stop operation of their magnox reactor (the KEDO project, which is now underway with backing from Japan and South Korea). Indeed, it is clear that there are many routes which lead to nuclear weapons: one almost universally used by the nine nuclear weapons states (five official, plus Israel, India, Pakistan and formerly South Africa) was uranium enrichment. The other was using plutonium produced in graphite and heavy water-moderated reactors which were dedicated for plutonium production with sometime, a secondary use for research or marginal power production. None of these countries used light water reactors to obtain their weapons material. Indeed, the power reactor route with LWRs is probably the least convenient and least practicable way to obtain material suitable for nuclear weapons. Therefore, any restriction on LWRs to prevent nuclear weapons proliferation is a little like blocking a small hole in the upper structure of a sinking ship which has plenty of large holes in its lower part. One can, therefore, question why the U.S. insists on stopping the Russian contract with Iran, which aims at finishing the LWRs at

Bushehr. Is it really for non-proliferation reasons or rather for general reasons connected with the status of Iran?

Another Issue which Some Classify under the Non-Proliferation Title Is the Diversion of Fissile Materials or Nuclear Weapons

It seems to me that this is not really a non-proliferation issue, but rather an issue of terrorism which should be dealt with in a similar way as the issue of diversion of some chemical or biological substances. Here clearly, very strict measures of physical protection should be developed and implemented. But one should also avoid a simplistic correlation between the quantity of separated plutonium and the risk of diversion. Indeed, the way plutonium is managed, in which country it is used, how far it travels, may be much more important than the quantity involved. In any case, the diversion of nuclear weapons themselves or fissile materials from nuclear weapons programs in some countries, notably those of the former Soviet Union, pose in my view a much larger threat today than the diversion of plutonium coming from the reprocessing of commercial reactor fuel Here again, one should first address the large holes before trying to block the small pinholes.

A More Controversial Issue Is the Assertion that the Spread of Nuclear Weapons to New States Is Always Catastrophic

Indeed, when new countries like the U.K. France or China became nuclear powers, countries which were already members of the "club" were strongly opposed, but now it is quite well accepted that possession of nuclear weapons by those countries did not provoke a catastrophe. Some even consider that the existence of nuclear weapons in different camps was a stabilizing factor during the Cold War and prevented a major conflict during the past half-century. Why, then, the possession by India should not stabilize the relation between India and China, and by Pakistan the relation between Pakistan and India, preventing major conflicts in these zones? At least the question may be asked. In the same way, the possession by Israel of nuclear weapons, in the opinion of some, has stabilized the situation in the Middle East. In any case, when a country has decided that it is worth while to make the effort and take the risks of developing nuclear weapons, it seems that after some initial outcry, the world accepts it without major retaliation. That shows some kind of hypocrisy in the initial claim of a fundamental evil connected with the proliferation of nuclear weapons. The real issue is probably to avoid the acquisition of nuclear weapons by unstable, not very democratic countries. May be Pakistan is in that category; clearly the West would not like Libya or Iran and Iraq to possess such weapons. A real, major issue is full nuclear disarmament, but this is another story.

COMPETITIVITY OF NUCLEAR POWER

Present Situation

It seems increasingly clear that the operating cost (operation and maintenance plus fuel costs) is very competitive with the same cost of its main competitors, coal-and gas fired plants. Therefore, the economic advantage of life extension is obvious, and also the clear trend to operate existing nuclear power plants on base load only. Even in France where there are too many nuclear plants to operate them all on base load, there is a trend to decrease the share of nuclear power in total production in view to operate eventually all nuclear plant on base load. The construction of new nuclear plants is, however, much more questionable. Under the most favorable circumstances, a series of four to 10 new standardized units built in France, where the regulatory environment is relatively stable-nuclear may produce a kilowatt-hour at about the same cost as that produced by combined-cycle gas fired plants or modern coal-fired plants, provided one compares baseload operation for at least a 30-year plant lifetime, assuming an 8% annual discount rate and gas prices slightly increasing in the future. Under these conditions, considering the low investment cost of gas plants and the possibility of building smaller gas-fired plants without a large economic penalty, the likely decision would be to build gas-fired plants. We should, however, mention that this does not include potential future developments regarding inclusion of external costs in the calculated cost of power production.

Possible Evolution of Competitivity

Global warming. Global warming, which is now seriously taken into account even by governments under the pressure of environmental groups and the general public, may have an important impact on energy production by leading to direct or indirect taxation (the latter in the context of tradable emissions permits) the emission of greenhouses gases. However, from the point of view of nuclear energy, there are still two questions marks: will taxes penalizing emissions of greenhouse gases be effective, and when? And how will that penalization influence nuclear? Indeed, even if in theory nuclear power should be exempt of any penalty, the same public opinion and the same pressure groups which tend to promote the anti-greenhouse taxes are not favorable today toward nuclear power. Therefore, they may try to oppose the logical application of potential penalties that might favor nuclear and try to promote a more general penalization on energy consumption.

The issue of progressive exhaustion of some fuel resources, notably oil and gas. The risk calculation for the construction of oil-and gas-fired power plants includes assumptions on the future prices of these commodities. It is, however, difficult, to assess the size of risk of price increase especially in light of projections, notably by the International Energy Agency, showing that a decrease in the quantity of oil produced will occur within the next 20 years and in the production of gas one or two decades later. Indeed, should this projection come true, the position of oil--and gas-producing countries will be much stronger and they may use that position to increase prices. One must ask whether the recent volatility of oil prices- more than a factor of two in one year- is just an accident or presents a new trend. In addition, over the longer term, the inevitable exhaustion of oil and gas may in the future be perceived by some ecologically oriented groups as a modification of the environment which is as important for mankind as the greenhouse effect, and it may happen on the same timescale as global warming-decades, not centuries. If this is the case, these groups may also apply pressure in view to save finite resources of oil and gas. That attitude would certainly be favorable to renewable energies; it might also offset a negative bias against nuclear energy.

Nuclear waste disposal. The resolution of the socio-political issue of nuclear waste disposal, necessary for the development of nuclear power, may appear more difficult than it is from the purely scientific or engineering viewpoint. However, this issue should be resolved not only for future plants, but also for existing plants which are operating now.

Regulatory requirements. Another issue which may make nuclear competitivity more difficult is a trend with some regulators to always improve the safety of new plants, even if the general safety of nuclear seems to be much better, at least in western design plants, than the safety of other means of producing electricity. Therefore, it is interesting to observe the trend in the U.S Nuclear Regulator Commission to maintain safety. This seems much more realistic, but it is not shared, for example, by French regulators, who may be under more pressure from parts of the French government.

CONCLUSION

We should add that the above remarks on future competitivity concern global trends. A more accurate appraisal should take into account the situation country by country, since situations may differ greatly.

TURKEY AND ENERGY SECURITY IN THE CAUCASUS AND CENTRAL ASIA

Paul Michael Wihbey

Strategic Fellow, The Institute for Advanced Strategic and Political Studies, Washington DC.
Global Foundation International Energy Forum, Banquet Address, November 1999, Washington DC.

INTRODUCTION

As the global system lurches from the old order of the Cold War into a new and yet undefined system of global power there are increasing signs of an emerging pan-Eurasian, anti-NATO alliance system between Russia and China. Although not yet a formal military alliance, Russian President Boris Yeltsin has already called this relationship "a strategic partnership."

Both Russia and China's worldview is based on a multi-polar balance of power aimed at countering U.S. global dominance. Both countries have a broad set of converging strategic interests which include; (1) a stable and secular central Asia free of Islamic political regimes; (2) maintenance of the primacy of the UN Security Council; (3) opposition to U.S-driven theater missile defense systems; (4) opposition to NATO-type intervention in sovereign states on behalf of human rights and minority rights; (5) the perception of NATO as no longer just a defensive alliance (as in the Kosovo crisis).

Central Asia

To this end on August 25, Yeltsin and Chinese leader Jiang Zemin along with the leaders of Kyrgystan, Kazakstan, and Tajikstan signed the Bishkek Declaration thereby enshrining some of the basic tenets of the new Russian/Chinese strategic alignment including cooperation on security issues, border control arrangements, and the affirmation of the principles of non-intervention and respect for national sovereignty.

The Challenges to Nuclear Power in the Twenty-First Century
Edited by Kursunoglu *et al.*, Kluwer Academic/Plenum Publishers, New York, 2000

As a result, the preconditions for a formal strategic alliance between Russia, China, Kyrgystan, Kazakstan, and Tajikistan have now, effectively been put in place. Such an alliance system could also attract a third nuclear power--India (i.e. Primakov's 'Strategic Triangle' gambit), as well as Uzbekistan. Uzebekistan, which has been pro-NATO, and a non-CIS security member has come under increasing internal stress because of an upsurge in Islamic insurgency and rebel activities. The distinct possibility now exists for the integration of the largest nuclear arsenal with the largest conventional army thereby achieving status of a military superpower with massive geostrategic capacity on the Eurasian landmass. Already, under the strategic planning body entitled, Intergovernmental Commission on Military and Technical Cooperation, co-chaired by the Russian First Deputy Prime Minister and the Deputy Chairman of the Chinese Central Military Commission, Russia and China have made significant progress on several fronts including;

- Compatibility of weapons systems,
- Sharing of intelligence,
- Increasing economic ties,
- Transfer of technology, and
- Arms sales, which could include the Russian SSN-22 supersonic anti-ship cruise missile, the SU-30 fighter, and Typhoon class nuclear ballistic submarines.

Complimenting this development, even historic development, has been the gradual reemergence of the CIS Collective Security community, which is currently composed of Russia, Belarus, Kyrgystan, Kazakstan, Tajikstan, and Armenia. Within the last month, both Uzbekistan and Kyrgyztan have sought Russian security and military assistance against Islamic insurgents. These actions undermine the NATO-driven "Partnership for Peace Program" that was so evident during NATO's 50th anniversary celebrations. Such requests simply highlight the *emerging issue-preeminence of regional security needs* over economic development and political reform, and the need for Central Asian states to seek security guarantees from Russia.

In conclusion, as regards Central Asia, I believe Russia, with Chinese support, will take advantage of an emerging security vacuum to extend its influence back into the Southern Eurasian heartland. Other contributing factors to this strategic impulse, I suggest, involve the perception of neo-isolationist tendencies in the United States; an American unwillingness to accept further deployments (i.e. East Timor); a risk-averse U.S. Administration reluctant to take further foreign policy initiatives during the Presidential electoral-cycle, and; Moscow's exploitation of domestic Russian popular sentiment demanding retribution for recent terrorist bombings, and, possibly, the bombing campaign against Serbia.

The Caucasus

This is a region of critical importance. Not withstanding the potential for conflict in the Pacific Rim, the Balkans, and between Israel and the Arabs, the Southern Caucasus/Northern Mesopotamia region is probably the most geo-strategically important piece of real estate in the world.

Indeed, if a radius of 300 miles is drawn at the point of convergence of the Iranian, Turkish and Armenian borders, just south of Yerevan, the territory thusly covered would include the following list of cross-border and internal conflicts that have occurred during the 1990's;

Cross Border Conflicts.	U.S. (multinational) versus Iraq, Armenia versus Azerbaijan
Major Insurgencies.	PKK versus Turkey, Chechen versus Russia, Abkazia versus Georgia, Iraqi opposition (INC) versus Iraq (Saddam Hussein).
As well as three other flash points.	Turkey /Syria; Turkey / Iran; and Azerbaijan/ Iran.

The situation is further compounded by major oil and gas deposits in Northern Iraq and Azerbaijan (Caspian Sea), as well as critically important and contentious pipeline routes such as Baku-Ceyhan and Baku-Supsa. The volatility of this region can be seen even today, with the deployment of Russian troops in Chechnya. Indeed the three states that make up the geostrategic unit of the South Caucasus, -- Georgia, Armenia and Azerbaijan, two of whom share borders with Turkey, will be directly impacted by the advancement of Russian forces in Dagestan and Chechnya.

A case in point is Georgia, a founding member of the western-oriented GUUAM pact (Georgia, Ukraine, Uzbekistan, Azerbaijan, and Moldava) for economic development. Moscow, at this time, seems intent on isolating the Shevardnadze administration by maintaining Russian troops and bases on Georgian territory, thereby reducing Tbilisi's capacity to deal with the Abkazia insurgency. Straddling the transit routes to the Supsa and Ceyhan oil terminals, Georgia is key to regional stability and prosperity. Not only is Georgia a strategic partner with Azerbaijan, but it maintains friendly and cordial relations with Armenia and Turkey. Consequently, any destabilization of Georgia would act as an incentive for the extension of the Russian sphere of influence to the very borders of Turkey. Combining such an eventuality with the already large presence of thousands of Russian troops in Armenia and Russian control over the North Caucasus leaves Azerbaijan vulnerable. Already flanked on its southern rim, by Russian ally Iran, Azerbaijan would find itself exposed and isolated, and susceptible to various forms of diplomatic pressure and extortion particularly over issues of energy distribution and production and development.

A Georgian collapse creates the necessary conditions for a dramatic shift in the regional balance of power that under a *worst case scenario* would probably be characterized by; 1) a PKK destabilization of southeastern Turkey (see Mehmet Ali Birand, *Posta Newspaper*, Sept. 17; wherein Russian authorities warned Turkey not to assist Chechen rebels or risk the resumption of PKK attacks), which would then, 2) facilitate the development of an air and land corridor connecting the Russian forward deployment in the South Caucasus with northern Iraq, thereby; 3) extending Russian influence further into the Persian Gulf and the Middle East. Such a future could disrupt or entirely cut off the East-West energy transit corridor concepts that have been promoted by the United States and which stretch from the central Asian states through the Caucasus via Turkey into Europe.

Clearly, there are different future scenarios for the region, but the manner in which Georgia can withstand the current stress arising from Russian geopolitical ambitions, and the manner by which the West assists or does not assist Georgia, I believe, will be a determinative indicator as to what type of future we may expect in this region, along with obvious national security implications for regional powers like Turkey and Israel.

Turkey's Role

Within this context, the question arises as to Turkey's role in the Caucasus and Central Asia. I believe for many reasons that you are already familiar with, Turkey is not only the gate for the West into these regions, but Turkey is the indispensable power. The Central Asian and South Caucasus states do not have any great desire, so far as I can ascertain, to fall under Russian dominance, a Pan-Eurasian strategic alliance, or succumb to Islamic theocracy.

But conditions today are not those of 1992 or 1996. These regions need to be elevated to a priority status by Western policy planners. Turkey can be the key stabilizer if it is given the appropriate diplomatic, political and financial support by countries like the U.S., Israel and Germany. Such a Turkish-led engagement needs to be part of an integrated and systematic effort at regional stability based on <u>balance of power</u> calculations. For your consideration and within the context of public policy debate, I would like to propose, the following four proposals leading to the goal of Central Asian and Southern Caucasus peace and stability;

Turkish Model. Continued and upgraded promotion of the Turkish political and economic model of a secular, multi-party, market democracy. Turkish and U.S-based NGO's, policy and educational institutes, and multilateral organizations like GUUAM and OSCE can play an important role in adapting the Turkish experience to the particular socio-economic and political conditions in selected countries. The focus of such effort ought to be Azerbaijan, Uzbekistan, and Turkmenistan.

South Caucasus Cooperation Council. Turkey could play an important and pivotal role in maintaining the integrity of the South Caucasus by hosting (or co-hosting with the U.S.) the leaders of Azerbaijan, Armenia and Georgia, through the auspices of the South Caucasus Cooperation Council.

Security Assistance. A Turkish-lead, U.S-supported program to provide security assistance to various Central Asian states, would certainly offer an alternative to the Russian option, or the politically unacceptable idea of direct U.S. engagement. Turkish experience in combating terrorism and insurgency, training and use of special forces in mountainous terrain and urban areas, and adept use of attack helicopters could in a relatively short time change the dynamics of actual and potential conflict in favor of the governments in countries such as Uzbekistan.

Energy. The Trans-Caspian pipeline project is the basis of collaboration between Turkey and the U.S. in the region. This integrated project has two parts, one being the

Baku-Ceyhan and the other the Turkmenistan/Turkey/Europe natural gas pipeline. This project should be regarded with the highest strategic consideration, not only for Turkey, which is a major energy consumer, but for surrounding countries whose demand for energy is based as much on security of supply as pricing. These pipeline projects are the late 20th century's equivalent of the great transcontinental railway systems that secured nationhood for countries like the United States and Canada in the late 19th century. Continuing delay on the decisions relating to the East-West energy corridor only contributes to reducing the chances of such projects ever being built. Although risk can be calculated and reduced, it can never be totally eliminated. Competitors are emerging and the market is in a constant state of flux. Baku-Ceyhan's importance is multi-dimensional with positive impact on regional security, economic development, and political stability.

This is a pipeline project whose time has come. Whatever is required to effect Baku-Ceyhan as the Main Export Pipeline for Caspian oil must be done, and done as soon as possible.

INDEX

GPSR Compliance

The European Union's (EU) General Product Safety Regulation (GPSR) is a set of rules that requires consumer products to be safe and our obligations to ensure this.

If you have any concerns about our products, you can contact us on ProductSafety@springernature.com

In case Publisher is established outside the EU, the EU authorized representative is:

Springer Nature Customer Service Center GmbH
Europaplatz 3
69115 Heidelberg, Germany

Batch number: 10837782

Printed by Printforce, the Netherlands